PLANET HUNTERS

The search for
extraterrestrial life

HIGH LIFE HIGHLAND	
3800 17 0027332 3	
Askews & Holts	Jul-2017
523.24	£16.95

REAKTION
BOOKS

Published by Reaktion Books Ltd
Unit 32, Waterside
44–48 Wharf Road
London N1 7UX, UK
www.reaktionbooks.co.uk

First published by Reaktion Books in 2017

English translation by Andy Brown
Translation © Reaktion Books 2017

Illustrations by Lucas Ellerbroek,
Brechje Schreinemakers and Koen Maaskant

Planetenjagers © 2014 by Lucas Ellerbroek
www.planetenjagers.nl
Originally published by Uitgeverij Prometheus, Amsterdam

The publisher gratefully acknowledges the support
of the Dutch Foundation for Literature

Printed and bound in Great Britain
by TJ International, Padstow, Cornwall

A catalogue record for this book is available
from the British Library

ISBN 978 1 78023 814 2

For my parents

Nature's imagination far surpasses our own
Richard P. Feynman

Planet Hunters' Timeline

CONTENTS

INTRODUCTION: NEW MARBLES 9

1 THE CENTURY OF BRUNO 16

2 THE LITTLE SAND RECKONER 39

3 AN INQUISITIVE MIND AND DEFECTIVE SIGHT 60

4 THE ORDER OF THE DOLPHIN 81

5 THE TENACIOUS INVENTOR 102

6 A PLANET IN PEGASUS 126

7 THE HUT IN THE CAR PARK 155

8 GOLDILOCKS AND THE RED DWARFS 175

9 BEER IN SPACE 194

10 THE SPACE REBELS 210

11 THE PLANETARY CIRCUS 231

GLOSSARY OF TERMS 242

REFERENCES 245

BIBLIOGRAPHY 250

ACKNOWLEDGEMENTS 258

INDEX 260

INTRODUCTION:
NEW MARBLES

A T THE AGE of eight, I had far from outgrown the 'why' phase. Why did dinosaurs become extinct? Why can an aeroplane fly? Why couldn't I stay up as long as my parents in the evenings? My curiosity was only encouraged by my grandfather, who was anything but an old-fashioned, grey-haired old man. It was he who first introduced me to the computer, one he had put together himself and which I spent many hours playing around with. Together, we also made drawings, wooden planes and little clay models. He often answered my questions for me or showed me something new. An eight-year-old boy is easily distracted, and especially if his grandfather has a – self-built! – computer with Tetris on it.

One Friday, my grandfather took me to the Teylers Museum in Haarlem. Full of scientific instruments and complicated machines, it was a paradise for grandpas and their grandchildren. We walked along the Spaarne, beside the canal, to the stately entrance topped with its bronze statues of the muses of art and science. Between them stands Nike, the winged goddess of victory, who is presenting them both with a crown of laurels.

As soon as we went in, I ran to the room with the prehistoric animals. At that time, dinosaurs took up an alarming proportion of my time and energy. This was just the place for me. I gaped at the collection of bones and skulls: a crocodile-like marine dinosaur, a rhino with two horns, a giant guinea pig that made me look with new respect at its two minuscule descendants that I so neglected at home. After the prehistoric animals, I went to the room with old scientific measuring instruments. I imagined that, in the evenings, after all the visitors had left, an absent-minded professor would sit here in silence and conduct experiments. I considered hiding myself somewhere so I could watch him.

My grandfather had more to show me. He took me to a large, oval room full of telescopes and other astronomical instruments. On a sort of small table, there was a gold-coloured ball from which seven rods of different lengths protruded. At the end of each rod, there was a porcelain marble. Some of the marbles also had rods sticking out of them, with even smaller marbles on their ends. One of the larger marbles was surrounded by a flat ring and had eight smaller marbles around it. On the side of the table, there was a lever that could be used to set the whole thing in motion via a system of cogwheels.

It was an orrery, my grandfather told me, a scale model of our solar system. The golden ball was the Sun, the marbles were the planets and the smallest marbles were moons. The Sun was a star, he said, just like the many little points of light we could see in the sky at night. The planets moved continually around the Sun, and the moons around the planets. The third marble from the Sun was our own planet, the Earth.

The orrery raised many questions for me. Why did the Earth only have one moon? I thought it was unfair that the other planets had more than we did. And, if the Sun was a star, why were the other stars so small? Why were there only seven planets in the orrery? I was sure I had heard that there were nine planets in the solar system. Even my grandfather couldn't answer all my questions.

'Grandpa,' I asked. 'Are these the only planets that exist?'

THE ORRERY was built in 1790 by instrument-maker George Adams for his namesake King George III of England. The king owned a gigantic collection of scientific instruments. According to his critics, he devoted more energy to science than to politics. It was during his reign that England lost its most important colony, America. In George III's time, the solar system was believed to consist of seven planets: Mercury, Venus, Earth, Mars, Jupiter, Saturn and Uranus. The first six had been known for centuries, as they are visible with the naked eye. The seventh, Uranus, was discovered in 1781 by the astronomer William Herschel with a large wooden telescope, which had also been built by order of George III. The telescope was exhibited some two hundred years later in Teylers Museum, in the same room as the orrery.

Since the orrery was built, two newcomers had been added to the list of known planets: Neptune and Pluto.[1] After that, no new planets had been discovered. So my grandfather could do nothing else than answer my question in the affirmative. Back then, in the early 1990s, the planets in our solar system were the only ones we knew of. The only planets in the universe.

ON A clear night, once it is dark enough, the stars are breathtaking. If you look closely, you can see a white band going right across the night sky. The distinctive white glow is caused by hundreds of millions of stars that are too far away for us to distinguish individually.

Some people find the stars romantic, some become aware of their own insignificance when confronted with the vastness of the universe, while others wonder if *Dancing with the Stars* is on the television in the comfort of their living rooms. But there is one question that has occurred to everyone who looks up at the stars at some time: are we alone in the universe?

The obvious answer would appear to be 'No'. The universe is so immeasurably large – why should we be unique? If the Sun is a star, why shouldn't all the other stars have planets, too? Might there not even be another planet like the Earth? A planet that, like the Earth, is teeming with life? Surely we can't be living in the only spot in the universe where life exists? American comedian and armchair philosopher George Carlin found the idea that we are alone in the universe a little too much to bear. 'If it's true that our species is alone in the universe,' he joked, 'then I'd have to say the universe aimed rather low and settled for very little.'

WHEN MY grandfather showed me round the Teylers Museum in the early 1990s, the solar system may still have been alone in the universe but, in people's imaginations, that universe had been full of life for a long time. For hundreds of years, writers, artists and other creative individuals had fantasized about what aliens from other planets might look like. From the exciting radio drama *War of the Worlds* to the comic *Men in Black* films, extra-terrestrial life has been a recurring theme throughout the ages and in a wide variety of art forms. The search for it, though, took place for a long time mainly on the silver screen.

But, before long, that was to change radically. In those same years, the early 1990s, a handful of people spread around the world were engaged on the same lonely mission: to find a planet around another star. That discovery would signify a real step forward in the search for extraterrestrial life: there could be no aliens without planets for them to live on. It was a motley group of people: an American engineer with a hopeless plan for a space satellite, a young Swiss student who was selecting stars to study for his PhD research, an astronomer and do-it-yourself expert in the Rocky Mountains looking for an excuse to build a telescope, a professor who ignored the disapproving shakes of the head of his colleagues to search for new planets, and a small organization that used radio dishes to try and pick up signals from aliens. In the early 1990s these people were starting to get a little less lonely. They met up with each other more and more at conferences where they discussed the probability of alien life and how we could ever discover it.

Their long search finally bore fruit. The first exoplanet – a planet orbiting a different star than the Sun – was discovered in 1995. In the years that followed, more and more astronomers joined the small group of pioneers. A second planet was found and, shortly after, a third. In no time, the floodgates were open.

In the past twenty years, the existence of nearly 3,000 exoplanets has been confirmed by different, independent measurements and some 2,500 'candidates' are still awaiting confirmation. And these figures are already out of date: since 2010 an average of four new exoplanets have been discovered every week. That means thousands of new planets in the firmament, thousands of new marbles around different golden balls in the oval room of Teylers Museum.

JUST LIKE Lego, football and my coin collection, astronomy played a recurring role in my life. It was a kind of on-off relationship: I had periods when I was very deeply interested in the universe and I studied books full of facts about planets, space travel and far-off galaxies. But it was not until the winter of 2009, a week after I graduated, that I first considered astronomy as a career. In an exploratory conversation with a professor from Amsterdam, he told me almost as an aside that he still had a space free for a PhD student. It would entail going to the Atacama Desert in Chile a number of times to observe stars with the Very Large Telescope, the largest telescope in the world (with the least original name in the world). That set me thinking, and a month later I was on a plane to Santiago. For four years, I conducted research into the origin of stars, which was indirectly related to the origin of planets. I heard a lot about the most exciting current development in astronomy: exoplanets were the talk of the town.

The study of exoplanets is the fastest growing discipline in the field of astronomy. Astronomers are leaving their old areas of research in droves to search for new planets. Young researchers are devoting themselves to new questions: what kind of planets exist out there? How many planets does the average star have? What planets are suitable for life? How can we ever prove that such life exists?

Before the 1990s, the world 'planet' had somewhat stuffy, old-fashioned connotations in astronomy. But these days it is as popular as the royal family in the gossip pages. Today, finders of new planets know that the eyes of the world are upon them. Every new discovery is reported in detail in the media, and brings us a small step closer to answering the question that is on everyone's lips: are we alone?

In the autumn of 2013 I travelled through the United States to present my own research at a number of universities. In every town I stopped at, I talked to planetary researchers. Some were veterans, pioneers who went against the flow and now work at the most prestigious institutions in the world. Others were ambitious young researchers who use innovative methods to explore the issue of extraterrestrial life.

The search for exoplanets started in the u.s. and Europe, but now researchers are cropping up everywhere, from Australia to Japan and from Chile to South Africa. And the discipline is no longer limited to astronomers: more and more chemists, biologists, geophysicists and other specialists are being bitten by the extraterrestrial bug. Slowly, I feel an increasing awareness that something big is about to happen. That the collective efforts of this colourful and ever-growing group will unearth the answer to the question of whether extraterrestrial life exists. And even if they do not find that answer, every planetary researcher will agree with the words of anthropologist Claude Lévi-Strauss: 'The wise man doesn't give the right answers, he poses the right questions.'

The question of whether extraterrestrial life exists is so old that it is no longer possible to trace who asked it first. This story starts in the Renaissance, the time when the scientific method was born. Questions were no longer answered with words, but with observations and experiments. And not every question could be freely asked. The first person who dared to ask this one went up in flames. That man was the first planet hunter.

one

THE CENTURY OF BRUNO

O N THE Campo de' Fiori, as on many squares in Rome, there is a bronze statue. It is a monk wearing a habit and with his hands folded over a book. From under his hood, he looks sternly out across the flower stalls and the crowds of shoppers. On his pedestal, defaced by graffiti and pigeon droppings, there is a plaque with the following inscription:

A BRVNO

IL SECOLO DA LUI DIVINATO

QUI

DOVE IL ROGO ARSE

(To Bruno – the century predicted by him – here
where the fire burned)

One Saturday morning more than four hundred years ago, close to the spot where he now stands cast in bronze and watching the square, the monk was tied to a stake by armed guards. It was not busy on the square that morning. The market vendors were setting up their stalls – they still had to sell their wares, even on the day of an execution. He was gagged and hung by his feet, while the flames blazed around him. A small crowd watched as he was burned alive. Afterwards, the ash was gathered and thrown into the Tiber. The books the condemned man had written were blacklisted. It was 17 February 1600, the day that the heretic Giordano Bruno died by fire, the day that his ideas would be forgotten.

That, at least, was the hope of the priests of the Inquisition, who had sentenced him a week before for a whole list of controversial opinions and beliefs, mainly of a religious nature. One of the charges on the long list was 'claiming the existence of a plurality of worlds and their eternity'. Bruno said that the stars we see as small dots of light at night are in fact distant suns and – like our own Sun – they had planets orbiting them. Four centuries later, Bruno was proven to have been right. Many people still consider him the spiritual father of the exoplanets.

Bruno spent the seven years prior to his execution in a cell in Tor di Nona, a prison on the Tiber. From time to time, his inquisitors would ask him if he was ready to renounce his beliefs. His answer was always a steadfast 'No', despite the torture that invariably followed. According to an eyewitness, Bruno responded to his inevitable death sentence by saying: 'You may be more afraid to bring that sentence against me than I am to accept it.' He was not afraid of his judges; his principles were dearer to him than his life. For that reason, Bruno is still seen today as a champion of free speech.[2]

Bruno is often portrayed as a visionary, a martyr to science, and to astronomy in particular. At a recent dinner of the science faculty of the University of Amsterdam, an emeritus professor challenged his colleagues from the biology department, saying: 'You thought it was difficult to convince people about the theory of evolution? Darwin had it easy. Astronomy – now that's a dangerous profession!'

In reality, Bruno's ideas about planets ('other worlds') were only a small part of the charges that led to his gruesome execution. Catholic essayist Robert P. Lockwood recently put this image of Bruno as a 'martyr to science' into perspective. 'Giordano Bruno', he wrote, 'died from a massive ego, intellectual pretension, a singular dishonesty, an overactive libido, and for being a miscreant priest who allowed himself to be ordained when he didn't believe any essential truths of the faith. He's a walking billboard for the Inquisition.'

Burning someone alive is perhaps a somewhat drastic response to such objectionable qualities, but there may be a core of truth in Lockwood's words. Looking at the charges, it was primarily Bruno's provocative personality that was such a thorn in the side of the Church. It is therefore not entirely accurate to portray him as someone who died for science. Strictly speaking, he was not even an astronomer. That profession, as we know it, did not come into being until the discovery and widespread availability of the telescope, more than ten years after Bruno's death. But that makes his predictions even more impressive.

During his life Giordano Bruno won a lot of people over, but he also had a great talent for getting into arguments. After his childhood in the small village of Nola, in the shadow of Mount Vesuvius, he attended the seminary in Naples. It soon became clear that he had strong opinions, a quality that did not endear him to his teachers. After completing his training as a priest, he had to flee the monastery and the city after a book by Erasmus had been found in his cell. He became a sort of itinerant monk, turning up in cities all over Europe and leaving again just as suddenly, often after rubbing someone or other up the wrong way.

In the sixteenth century the adage 'knowledge is power' was rapidly gathering momentum. The rich elite of Europe became mesmerized by the pioneers of the Enlightenment, who propagated a new sort of science. In imitation of these groundbreakers, members of the nobility raised their status by learning facts, figures and pieces of text off by heart. Bruno cleverly took advantage of this trend and became renowned as a professional memory trainer. With his charm and powers of persuasion, he became a favourite of many prominent figures. Aristocrats throughout Europe engaged him to train their memories. This success made him no easier to get on with and he continued to live up to his reputation as a troublemaker.

Besides his writings on memory training, he published prolifically on his own 'Nolan' philosophy (named after his home village). He also taught at a number of leading universities, including Toulouse, the Sorbonne in Paris, and Oxford. Through his aristocratic connections, he had a wide readership and there are regular reports from the time of prominent figures discussing his ideas. Not everyone, however, was equally impressed: British academics found the quick-tempered little Italian more laughable than keen-witted. One eyewitness described Bruno raving on 'every conceivable subject' in a debate with an Oxford theologian. In his own account of the discussion, Bruno called his opponent a 'rough and rude pig' while he himself, as a real Neapolitan, was a paragon of 'patience and humanity'. His series of lectures in Oxford came to a premature end.

Such performances made Bruno less than popular with the magisterium of the Catholic Church, to put it mildly. One thing they particularly objected to was Bruno's use of the term *asino* (ass) for everyone he considered short-sighted. Furthermore, his Nolan philosophy was packed full with blasphemies, aimed especially at Jesus and Mary. Bruno doubted the divinity of the son and the virginity of the mother. He also refused to write in Latin, the language of the Church, preferring Italian, the language of the people. He was excommunicated twice, by both the Catholic and the Reformed churches. In short, he was high up on the wanted list of the Inquisition in Rome.

A conflict with one of his rich employers was ultimately his downfall. Venetian patrician Giovanni Mocenigo was dissatisfied with Bruno's dedication as a memory trainer. It didn't help that Bruno was rumoured to have been overly attentive to Signora Mocenigo. One day, Bruno informed his host that he was planning to attend the Frankfurt Book Fair. Mocenigo most likely saw this as desertion and it seemed a good opportunity to be rid of his irritating guest once and for all. He betrayed Bruno to the Inquisition. That night, he was roughly removed from his bed and, eventually, shipped off to the cell in Rome where he would spend the final seven years of his life.

One of the many sins he had to deliberate on while in his prison cell was his claim that the universe was infinite. He had also stated that the Creation, and therefore the Earth, was not unique. If the Sun was a star and the Earth a planet, Bruno reasoned, why did the other stars not have planets? What made the Earth the centre of the cosmos?

BRUNO DERIVED these ideas partly from a book by Mikołaj Kopernik, better known today as Nicolaus Copernicus, a Polish scholar who died five years before Bruno was born. Copernicus was born in the city of Toruń in 1473, the son of a wealthy merchant. He lost his parents at an early age and was adopted by his uncle, a bishop. He studied first at the University of Kraków, later going to Bologna to study law and then to Padua for medicine. While in Italy, he developed an interest in astronomy.

In his career, Copernicus had not bet on only one horse. After completing his studies in Italy, he returned to Poland to be his uncle's personal physician. He also occupied himself with reforming the Prussian currency system, a kind of predecessor of the euro. Despite this being a busy job for a man of his background and standing, there was sufficient time left over for astronomy. That resulted in a unique perspective on the planets and the Sun. His work would lead to a scientific coup that would be named after him: the Copernican Revolution.

To understand the full significance of this revolution, a little background is required on the nature of planets. With the naked eye, planets cannot be distinguished from stars: they are clear points of light in the night sky. They do not emit light themselves, but reflect that of the Sun. They are only visible at night, when the Sun illuminates them from behind the Earth. The most prominent planets in the night sky are Jupiter and Venus. They are brighter than Sirius, the brightest star. Saturn is difficult to distinguish from the stars with the naked eye, but its rings can be seen clearly through a telescope. The red planet Mars is perhaps the most recognizable. Mercury, the smallest planet in the solar system, is a special case. It is rarely visible, but when it is, it is a great spectacle: a very bright planet low on the horizon, close to the rising or setting Sun. The most distant planets, shy Uranus and Neptune, cannot be seen with the naked eye and were not discovered until the late eighteenth and mid-nineteenth centuries respectively.

The status and divine symbolism of the planets is clear from them being named after Greek and Roman gods. In French and Italian, most days of the week are still named after the planets and the Moon (*lundi* is Moon-day, *mardi* is Mars-day, *mercredi* is Mercury-day, *jeudi* is Jupiter-day and *vendredi* is Venus-day). In English, only Sunday (the Sun), Monday (the Moon) – together with Saturday (Saturn) – are left; the other days have been hijacked by Germanic gods.

The word 'planet' comes from the Greek word πλανή της (*planètes*), meaning 'wanderer'. Despite their resemblance to the stars, the planets have always had a special status, because they appear in different parts of the sky from night to night. The stars, on the other hand, have fixed positions in the firmament. They move across the sky during the night, like the planets, but the distance between them remains the same. The plough in the constellation of the Great Bear and the three stars in Orion's belt are always recognizable. The planets, on the other hand, move in respect of the 'fixed' stars from day to day. A planet might be visible in the constellation of Sagittarius one night, only to be found on Scorpio on another. These shifting points of light used to be known as 'wandering stars'.

The planets do not, however, wander through the sky at random. If you accurately observe their positions every night, you will see that they all move between the stars along one fixed route. This slightly curved line passes through the constellations of the zodiac. The Sun moves along the same line every day. That suggests that the paths of the planets, the Sun and the Earth all lie on the same flat plane. Check it out: place a handful of marbles on a flat table and imagine that you live on one of them. Looking out from that marble, you can connect all the others with one line.

This is how the planets, the Earth and the Sun have been envisaged in every model of the cosmos devised throughout history. The first generally accepted model was created by the Greek philosopher Aristotle, who lived in the fourth century BC. Aristotle claimed that the Earth was standing still: after all, there was nothing to suggest that it was moving. That resulted in his geocentric model, with a static Earth at its centre. The Sun and the planets moved around it like a carousel, on fixed circular orbits on a flat plane. This whole system was contained within an outer shell, a kind of disco ball to which the static stars were fixed.

Aristotle noted one peculiar feature of the orbits of the planets: if he tracked the orbit of Mars, for example, from day to day, he observed that it sometimes changed direction. It would pass through a constellation, turn around as though it had forgotten something, and then continue its journey in the original direction. This retrograde motion excluded the possibility that the planets moved around the Earth in perfect circles.

Aristotle came up with a clever trick to explain this phenomenon: besides orbiting the Earth, the planets also made smaller circular movements. You can compare it with the Octopus, the 'spin-'n-puke' ride at the funfair. The Octopus has a number of arms that turn on a central axis, and cars at the end of the arms that also spin around. That means that each car follows a complicated path made up of larger and smaller circular movements (guaranteed to make you feel sick). Aristotle used these double circles to explain why, when seen from the central point (the Earth), the planets sometimes seemed to be moving back and forth. He called the smaller circular movements epicycles.

As measurements became increasingly accurate, more and more epicycles proved necessary to explain the movements of the planets. In the second century AD, the astronomer Ptolemy expanded Aristotle's model by adding many more epicycles on top of epicycles that turned in and around each other like cogwheels. Every planet moved in a circle, which in turn moved in a larger circle, and then sometimes in an even bigger circle. Although it looked a little messy and contrived, this new model described the movements of the planets with reasonable accuracy. A few centuries after it was published, Ptolemy's interpretation even received the official seal of the Church: it provided a satisfactory explanation of the movements we see in the night sky and, furthermore, the notion of the Earth as the centre of the cosmos was compatible with the teaching of the Bible. That visual confirmation and the approval of the religious authorities guaranteed Ptolemy's system a long life: for almost two thousand years, it was the only permitted view of the cosmos.

But Copernicus wasn't happy about it. When he was about forty, he presented his friends with an alternative model for the solar system. This model was radically different: the Sun, rather than the Earth, was at the centre of the cosmos, and the planets – including the Earth – moved around it. This heliocentric model was not entirely new. As we shall see, it had also been proposed in Greek Antiquity. Copernicus, however, furnished it with a scientific underpinning which explained the back and forth movements of the planets in much simpler terms. Instead of epicycles, he claimed that these movements were caused by the planets 'overtaking' each other. If the Earth and Mars orbit the Sun at different speeds, at the moment when the Earth overtakes Mars on the 'inside bend', we will see the red planet move back and forth in the sky.[3]

Copernicus' model made the solar system a lot simpler and – for some – more acceptable. He was, however, aware of the impact his ideas would have if he were to publish them, saying 'the contempt which I had to fear because of the novelty and apparent absurdity of my view, nearly induced me to abandon utterly the work I had begun.' Despite being urged to do so by his friends and colleagues, Copernicus did not make the model publicly known. Only at a very advanced age did he allow himself to be persuaded by his less cautious pupil Rheticus. Around his seventieth birthday, Copernicus described his model in detail in the book *De Revolutionibus orbium coelestium* (On the Revolutions of the Heavenly Spheres). It was one of his last deeds: the first copy of his book reached him in 1543 on his deathbed, only hours before he died. In a most likely dramatized version of this event, the old scholar awoke from a coma, leafed through the book, and died with a smile on his face, aware of the uproar it would unleash.

Later, the title of the book proved to have a double meaning. *De Revolutionibus* is about the revolutions of planets, but it is also perhaps the most revolutionary book in the history of science. It changed the way people thought about the Earth: rather than being the centre of the universe, it was no more than a very small part of it. The impact attributed to it is comparable to that of Newton's *Principia* and Darwin's *The Origin of Species*. That makes it even more remarkable that so few copies of the book exist. In the four hundred years since it was first published, it has only been reprinted five times. The select group of readers included famous astronomers like Kepler and Galileo, who would eventually be jointly responsible for the Copernican Revolution.

Despite its title, the book was not originally written with the intention of unleashing a revolution. It contains no attacks on the established order, no solemn proclamation of a new world-view. In the preface, in fact, the reader is requested not to take the content too literally and to see it only as a mathematical hypothesis. It is a disclaimer, added by the publisher at the last moment to smooth the ruffled feathers of religious readers.

Copernicus' model was by no means constructed perfectly and logically. Most of the elements attributed to the Copernican Revolution cannot even be found in the book, but were added in later interpretations. The epicycles from Ptolemy's model had not yet disappeared completely. In fact, after the fine-tuning that Copernicus had to do afterwards to explain his observations, there were even more of them. In his model, the planets did not consist of the same material as the Earth. The Sun was not seen as a star, and the universe was small and orderly. Despite these limitations, the heliocentric model can be seen as a first, daring step in the right direction. It was the first mathematically underpinned description of the solar system with the Sun in the middle and a moving Earth. It was a courageous shot in the dark.

Copernicus had distributed his model to a small group of friends and acquaintances and, thanks to his caution, he did not suffer much from sceptics during his lifetime. His different view of the world seemed to draw the attention of opponents only on a few occasions. The first instance of mockery occurred around 1531, during the carnival celebrations in the Prussian city of Elbing, when a caricature of a star-divining Copernicus featured in a farce. The second was eight years later when, after reading a pre-publication version of *De Revolutionibus*, reformist Martin Luther was led to remark disapprovingly,

> There is talk of a new astrologer,[4] who wants to prove that the Earth moves and goes around instead of the sky, the Sun, the Moon, just as if somebody were moving in a carriage or ship might hold that he was sitting still and at rest while the earth and the trees walked and moved . . . The fool wants to turn the whole art of astronomy upside-down.

Copernicus' ideas also caused little consternation in the years after his death. There was, however, a development that would eventually make it possible to test their verity: the advent of the scientific experiment. At the end of the sixteenth century, the natural sciences were in the process of separating from philosophy. Before then, the two disciplines had been happily married. The Greek philosophers sought explanations for natural phenomena by thinking about them, not by measuring them. That changed with the invention of the microscope, the telescope and other measuring instruments. Science became a time-consuming affair; anyone who spent their time staring through an eyepiece had no time to philosophize. To measure became to know.

On the eve of this sea change, Giordano Bruno read an English commentary on *De Revolutionibus* in a library in Oxford. He took the daring step of up-scaling the heliocentric model to include the stars. That made him one of the first and most radical pioneers of the Copernican Revolution.

IN 1584 Bruno published a book with the ambitious title *De l'infinito, universo e mondi* (On the Infinite Universe and Worlds), in which there is a kind of Socratic dialogue between a teacher (Philotheo, a recurring character in his books with pedantic, Bruno-like characteristics) and a pupil (the rather sheepish Elpino). Philotheo has a tendency to answer every question with a question:

> ELPINO: How is it possible that the universe can be infinite?
> PHILOTHEO: How is it possible that the universe can be finite?
> ELPINO: Do you claim that you can demonstrate this infinitude?
> PHILOTHEO: Do you claim that you can demonstrate this finitude?
> ELPINO: What is this spreading forth?
> PHILOTHEO: What is this limit?

Rather than throw a chair at this teacher, which would have been a perfectly understandable and justified reaction to such a tiresome method of argument, Elpino becomes convinced – after a handful of questions back and forth – of the infinity of space. Teacher and pupil continue to philosophize about the implications of this conclusion. If the universe is infinite, why would the night sky be an immobile ceiling, as had been believed until then? Was it not more likely that the stars are just the same as our own Sun: spheres emitting light and floating through space, each following their own path? If this were true, then the stars would have to be very far away, since they are much smaller and fainter than the Sun. And it is because they are so far away that we don't see the stars whizzing past, but fixed in the sky.

Elpino comes to the conclusion that the stars are the same as the Sun and then takes this line of thinking a step further, a step that proves very prophetic: he applies the Copernican model of the solar system to the stars.

ELPINO: There are then innumerable suns, and an infinite number of earths revolve around those suns, just as the seven we can observe revolve around this sun which is close to us.[5]
PHILOTHEO: So it is.
ELPINO: Why then do we not see the other bright bodies which are earths circling around the bright bodies which are suns?
PHILOTHEO: The reason is that we discern only the largest suns, immense bodies. But we do not discern the earths because, being much smaller, they are invisible to us.

Bruno, speaking as Philotheo, puts his finger on a very tender spot. To this day, we have been unable to find a twin for the Earth orbiting another star, for the very reason he gives: the planets are so small. It is unbelievable that he states his arguments so clearly in an age when so little was known about the properties and distances of stars. For Bruno, it added a few extra logs to the fire that was to take his life. Today, it is a reason to give an ambitious planet hunter more money to buy a bigger telescope.

A significant first attempt to estimate the immense distance to the stars was made nearly a century after Bruno's death by fire, from a field in Delft, in the Netherlands. But before we come to that, we will first look at how Copernicus' ideas fared further.

Two weeks before Giordano Bruno's death, in a castle near Prague, a meeting took place that would change astronomy forever. It was the start of one of the most significant, famous and difficult working relationships in the history of science: between two men who – as their biographer Arthur Koestler put it – 'were opposites in every respect but one: the irritable, choleric disposition which they shared.' The two collaborators were Tycho Brahe and Johannes Kepler.

TYCHO BRAHE was a Danish aristocrat. As a young boy, he had seen a solar eclipse. The event had been precisely predicted many years in advance in astronomical tables. From that moment, Tycho was fascinated by the fact that you could predict the movements of celestial bodies to the very minute.

He became even more motivated when he discovered that some of the predictions of his predecessors were inaccurate. At the age of seventeen, he observed a conjunction of Saturn and Jupiter: the two planets were so close together that they were hardly distinguishable from each other. To his amazement, Tycho saw that the tables he consulted had predicted this event a month too late. It was a revelation. The tables had been compiled by respected astronomers and had an almost biblical authority. How could they have been so mistaken? Tycho realized – perhaps one of the first to do so – that astronomers could enormously improve their ability to predict by taking more frequent and more accurate observations. From that moment, he saw it as his sacred task to map the heavens as accurately as possible.

Details in the sky are measured in degrees, arc minutes and arc seconds. The diameter of a pea at arm's length is about 1 degree wide. There are 60 arc minutes in a degree, and 60 arc seconds in an arc minute. The smallest detail that can be distinguished by the human eye is approximately 4 arc minutes – the same pea at a distance of 10 metres.

Tycho succeeded in measuring the positions of stars and planets to an accuracy of 30 arc seconds. That is equivalent to the apparent size of a pea at the other end of a football field. He achieved this degree of precision, unprecedented at that time, without using lenses. His accurate instruments were comparable to a slide rule, with two measuring sticks sliding past each other. He also had the patience of a saint – an unexpected characteristic given his other vices.

Tycho was known for being hot-headed. While at university, he lost a large piece of his nose in a fencing duel with a fellow student. The reason for the duel was a matter of honour: who was the best mathematician. After that, Tycho wore an artificial nose made of copper. I'll leave it up to the reader to judge whether this improved his countenance. Tycho wanted to devote himself to science. There would be time enough later to waste his time 'like other nobles . . . on horses, dogs, and luxury', as he himself described it in the introduction to one of his books.

Tycho had no lack of luxury during his scientific career. If, as a modern-day astronomer, you get the opportunity to spend a night in an observatory, you are over the Moon. Tycho was given an island in the Øresund strait to build an observatory. The island was called Ven, and the palace and observatory that Tycho had built was called Uraniborg, the castle of Urania, the Greek muse of astronomy. It became a kind of temple of science, to the glorification of Tycho, the self-proclaimed guardian of astronomy. From the outside, it looked like a large iced cake, with pointed domes, decorous arches and rows of columns. The rooms bulged with complex instruments, the most impressive being the equatorial armillary sphere, a copper colossus three metres high and wide. Tycho could determine the position of a planet by looking along the measuring rod. Then two assistants would turn a large wheel, rotating the rod by 180 degrees, so that the master could take a second measurement within a minute. Repeating measurements was important and innovative, minimizing errors, for example because of irregularities in the measuring rod.

In the big room, there was a magnificent model of the solar system and a painting of Tycho, surrounded by his instruments and assistants, portrayed much smaller than their master. The local people saw this strange figure, with his complicated instruments, as a magician and called Ven 'the Sorcerer's Island'.

In Uraniborg, Tycho received his noble guests, who encountered the members of his colourful household. There was a dwarf called Jepp, who used to sit under the table during banquets and shout obscenities at the guests (a kind of Tyrion Lannister from the novel series *A Song of Ice and Fire*). Another notable member of Tycho's household was his tame elk. During one evening of excess, the pet beast drank too much beer, fell down the stairs and died. These days, dwarfs and elks play a slightly less prominent role in astronomy.

In the heyday of his time on Ven, Tycho squandered 10 per cent of Denmark's state expenditure, and not exclusively on scientific activities. Such a state of affairs could not last forever. After a binge lasting twenty years, he left the island. The reason was probably a combination of disagreement with the new Danish king and the boredom of someone who already has everything. After moving around for a few years, Tycho settled in a castle in the Bohemian village of Benátky, as court astronomer to emperor Rudolf II in nearby Prague.

Tycho's life of leisure is in stark contrast to the hardships suffered by Johannes Kepler. Kepler was born in Germany in 1571, in the village of Weil, near Stuttgart. His childhood was an unhappy one. Europe was one enormous battlefield, with almost every country at war with one or more of its neighbours. When young Johannes could still barely walk, his father left to fight the Dutch as a mercenary in the Spanish army. Those left behind at home had their hands full with one of Johannes' brothers, who suffered from epilepsy and psychosis. Their mother was a herbal healer who was later accused of witchcraft. Kepler had to fight for five years to save her from burning at the stake. Besides all these tribulations, he had a weak constitution and suffered from chronic rashes, short-sightedness and haemorrhoids. He made few friends at university. On paper, Kepler's exceptional talent for mathematics seemed the only ray of light in his life.

When Kepler received the invitation to become Tycho's assistant, it must have seemed like a gift from heaven. After he had been thrown out of his job as a mathematics teacher in Graz, Austria, because of his Calvinist leanings, he had ended up destitute in Prague.

The publication of his book *Mysterium cosmographicum* heralded the start of a somewhat more positive period in Kepler's life. The book built on Copernicus' ideas, presenting a helio-centric model of the solar system. The planets moved in circular orbits around the Sun and the distances between them were determined by shapes that fitted into each other. The orbit of Saturn, for example, fitted exactly around a cube, while that of Jupiter described a circle that fitted precisely within that cube. And within that circle was a triangle in which the orbit of Mars fitted. In Kepler's model, the solar system fitted together like a Russian matryoshka doll. The shapes were fine and elegant and the planetary orbits matched (almost) the measurements available at the time. Kepler was justifiably proud of his model, which demonstrated the wondrous symmetry of the universe.

Tycho had read the book and did not agree with Kepler. He had his own perspective on the universe. The model he ascribed to was a kind of religious compromise between Ptolemy and Copernicus: the planets orbited the Sun, but the Sun orbited the Earth, which retained its place at the centre of the uni-verse. At that time, Tycho was at loggerheads with his arch-rival Ursus, who had yet another model of the cosmos. In the pages of *Mysterium*, however, Tycho immediately recognized Kepler's mathematical talent. This time, he did not want to fight against that talent (he may have rubbed his false nose pensively), but make use of it for his own gain. He wanted to engage an assistant who could prove that his own model was right. After all, he made very accurate, new measurements and a brilliant mathematician like Kepler should be able to fiddle around with them enough to confirm Tycho's predictions. But things went wrong between the nobleman and the maths teacher right from the outset.

Arthur Koestler's wonderful book *The Sleepwalkers* describes the setting of their meeting. Tycho's household lay in ruins. His instruments had been badly transported from Ven. The plague had broken out. Several members of the castle staff had left after disagreements with their boss, or had simply not turned up. The head of the household himself was constantly in conflict with the local authorities, who refused to pay his extravagant fee. To make matters worse, his daughter was having an affair with one of his assistants. Nor did it help much that Kepler had insulted Tycho by sending Ursus a rather gushing letter shortly before their meeting. Tycho's sons, who picked Kepler up in Prague, behaved jealously and with contempt. In short, by the time Kepler crossed the moat to enter the castle in Benátky, his spirits must have been very low indeed.

For the next year and a half the two astronomers made each other's lives a misery. The extravagant and chaotic household drove the moderate Kepler half to distraction, so that he had to leave the castle on several occasions. He applied in vain for other positions. His work was made even more difficult because the paranoid Tycho refused to share his measurements with his assistant. This made it impossible for Kepler to perform the task for which he had been engaged.

Tycho's death in 1601, a year and a half after Kepler arrived in Prague, gave the latter's career a considerable boost. During a dinner, Tycho waited too long to urinate (according to Kepler, so as not to appear impolite), which resulted in a fatal bladder infection. This explanation has been challenged very recently after traces of mercury, which is very toxic, were found in Tycho's moustache hairs. Could his vindictive assistant have poisoned him? In 2010, Danish scientists exhumed Tycho's remains to answer this question. The tests showed that he had not been poisoned; the mercury was probably the remnant of his alchemical experiments and his artificial nose. So a bladder infection remained the accepted cause of death. The unruly Tycho proved to have table manners after all, which were to cost him his life.

YOU CAN divide astronomers into two kinds: observers, like Tycho, and theorists, like Kepler. An observer describes what he sees as accurately as possible. A theorist tries to find an explanation that will also help predict future developments – sort of like fortune-telling with evidence to back it up. Sometimes theories are drawn up before the predicted phenomenon has been observed. Albert Einstein's general theory of relativity is a good example. In the early twentieth century, Einstein predicted that light could be bent by the gravity of stars. His theory was only confirmed later by observation. Often, it works in the other direction and observation comes before the theory.

This was the case with Tycho and Kepler. The theorist Kepler succeeded Tycho, the classical observer, as imperial astronomer in Prague. He was given access to all of Tycho's measurements and instruments. He set himself the target of explaining all the measurements with a new theoretical model of the solar system, preferably of course his own, with its circles, squares and triangles all fitting into each other.

In the eleven years after Tycho's death, Kepler tried to calculate the orbits of Mars and the other planets. He discovered that his geometric model was not up to the job. Even after making exhaustive adjustments, he could not make the observations fit a system of circular orbits. Finally, he formulated what would later be known as his 'three laws of planetary motion' (he never actually formulated them as laws himself; he identified them as notable phenomena that followed from his new theory). He presented the third law in a different book to the first two, and later toned it down. He also claimed that the regularity of the relationship between planetary orbits was based on musical harmonies, a conclusion that also proved untenable.

To this day, scientists still wrestle with the dilemma between 'beauty' (natural theories must be above all neat and elegant and expressible in geometric forms and symmetries) and 'truth' (reality is often complicated, chaotic and difficult to arrange in orderly patterns). We hope that, behind the complex natural phenomenon that we observe, there is order and regularity. Einstein was convinced until his death in 1955 that the cosmos could be explained with a logical, neat and ultimately simple 'theory of everything'. Leon Lederman, another Nobel Prizewinner, described this goal in 1988 as 'a formula so elegant and simple it will fit easily on the front of a T-shirt'.

That T-shirt has not yet been printed, but Kepler's laws have been a standard part of the curriculum for first-year students of physics and astronomy for hundreds of years. They state that each planet orbits the Sun in an ellipse and that their orbital velocity is not constant. The third law predicts exactly the relationship between the orbital period and the orbital velocity of a planet. The further a planet is from the Sun, the slower it moves and the longer it takes to complete one orbit. Jupiter, which is five times further from the Sun than the Earth, takes almost twelve terrestrial years to complete one orbit.

Kepler's laws were not explained until more than half a century after his death. Isaac Newton stated that objects with mass attract each other with an invisible force which he called 'gravity'. An apple is attracted to the Earth by the same force that keeps the planets in their orbits around the Sun. Gravity decreases as the distance between objects increases; the further planets are from the Sun, the weaker the reciprocal attraction and the slower their orbital velocity, just as Kepler had predicted.

Kepler's genius is also illustrated by his willingness to admit that he was wrong. He sacrificed his own geometric matryoshka model, which he was so proud of, because it was not compatible with Tycho's observations. This made him an important pioneer of the modern scientific method, just like Galileo Galilei and the British philosopher Francis Bacon, who were seven and ten years older than Kepler, respectively. According to this method, observations and experiments were used to test theories. The theories are then adjusted and again tested with new experiments, and so on. Theories can therefore never be proved, only disproved. Those that are still valid after many rounds of experimentation are the most durable. After Kepler had abandoned his old model, he modestly emphasized in his publications that his new theories might one day be disproved.

The champions of this scientific revolution did not actively propagate it themselves. Copernicus kept his new view of the world to himself. Bruno's exoplanets were only one of the many criminal allegations he had to answer for. Tycho Brahe was primarily concerned with increasing his status as an astronomical cult figure. The pessimistic Johannes Kepler perpetually drew attention to the shortcomings of his brilliant ideas. Newton wrote much more about alchemy than gravity.[6] In a nutshell, the importance of scientific discoveries is often difficult to estimate at the time they are made.

WITH HINDSIGHT, we can see that the European Renaissance changed our view of the heavens. Until then, astronomy had largely been a matter of measurement: it went little further than describing the recurring movements of points of light in the sky. Kepler's laws, however, made the solar system into a logical mechanism, of which the Earth was only a small component.

Newton showed that celestial bodies moved according to the universal laws of nature, which also applied on the Earth. From the seventeenth century, this new world-view started slowly but surely to gain ground. That was largely due to a significant invention: the telescope.

The first example of a telescope-like instrument that we know of was constructed by spectacle-maker Johannes Lipperhey from Middelburg, in the Netherlands, who probably stole the idea from his neighbour and rival, Sacharias Jansen. Lipperhey applied for a patent in 1608 for a device he had made with two lenses, describing it as a tube 'for seeing things far away as if they were nearby'. The patent was not granted, but the 'Dutch perspective glass' soon became a popular instrument of war on Europe's battlefields.

It was an Italian mathematics professor (who, as chance would have it, had just stolen an appointment in Padua from under Giordano Bruno's nose) who saw that, besides its use on the battlefield, the new instrument had potential scientific value. Galileo Galilei made one to his own design and focused it on celestial bodies rather than enemy lines. Through the enlarged images, he was able to discern all kinds of details that had not previously been observed: craters on the Moon, Venus as a crescent half illuminated by the Sun and, to cap it all, four moons orbiting Jupiter. Galileo's discoveries reinforced his belief in a heliocentric universe. He had, after all, seen with his own eyes that not everything in the solar system rotated around the Earth and that the surfaces of celestial bodies closely resembled those of the Earth.

He reported on his discoveries in a pamphlet entitled *Sidereus Nuncius* or *The Starry Messenger*, which was widely read. In his *Dialogo dei due massimi sistemi del mondo* (Dialogue Concerning the Two Chief World Systems), Galileo juxtaposes the geocentric world-view (with the Earth at its centre) with his own heliocentric system. In the book, the geocentric model is defended by a Dominican monk suggestively called Simplicio. Not entirely by coincidence, the sitting Pope was from the same Dominican order. Galileo had many talents, but tact was not one of them.

It therefore hardly came as a surprise when he was called to order by Bellarmine, the same cardinal who had been responsible for torturing and sentencing Bruno some years earlier. The story goes that Bellarmine took Galileo to the dungeons of Tor di Nona and showed him the instruments used to torture Bruno, at which the professor withdrew his claims. He spent the rest of his life under house arrest. He was blind when he died, caused by looking at the Sun too often through his telescope, something that – in the words of an old astronomical joke – you do only twice: once with your left eye and once with your right. 'EPPUR SI MUOVE' ('And yet it moves') became Galileo's defiant epitaph, referring to the movement of the Earth around the Sun.

AFTER KEPLER learned of the discovery of Jupiter's moons, he bombarded Galileo – or, as he called him, 'the Italian whose last name is the same as his first' – with fan mail. Kepler did not think much of Bruno's idea of an infinite universe, with an infinite number of stars and planets. He had written before of the insanity among philosophers to extrapolate the Copernican model to the other stars. He did not consider 'wandering through infinity' a good thing. 'You have restored me not a little by your labours,' he wrote to Galileo. 'If you had found planets circling one of the fixed stars, there among Bruno's infinities, I had already prepared my prison shackles, that is, my exile in that Infinity. Thus you freed me from the great fear that I had conceived when I first heard about your book.'

As many as four centuries after Kepler and Galileo, people still widely believed there was only one solar system. Most scientists, including Kepler and Copernicus, were convinced that there was a certain aesthetic, divine design behind the universe.

That could have been the work of a Creator in the literal sense of the word, or of a kind of mathematical God who spoke to us through nature and its symmetries. That is why explanations were sought in geometric forms or, as in Kepler's later model, musical harmony. Countless planets wildly circling other stars and disrupting that harmony was a frightening notion for Kepler.

We now know of the existence of hundreds of other solar systems, each with a completely different design. Planets come in all shapes and sizes, with orbits of all conceivable lengths and variations. It is ironic that the space telescope that found most of these hundreds of new planets was named after Kepler, the man who was afraid of exoplanets. Many of my Italian colleagues feel that the telescope should actually have been named after the real spiritual father of exoplanets, the rebellious dreamer Giordano Bruno.

Let us return to the Campo de' Fiori and re-read the text that appears on his statue, but now from a different perspective: 'FOR BRUNO. THE CENTURY PREDICTED BY HIM. HERE WHERE THE FIRE BURNED.' Bruno's many worlds have at last been found. The 'century he predicted' has finally dawned. It has taken more than four hundred years for his prediction to come true. Bruno himself gave the reason for this, in the words of the pedant Philotheo: the stars are so far away that their planets can hardly be seen. Just how far away, we would soon discover.

two

THE LITTLE SAND RECKONER

SOMETIMES OUR curiosity is overpowered by our inability to comprehend large numbers. I am drinking a cup of coffee with a retired astronomer in Amsterdam. He tells me that, the previous Sunday, his neighbour had come around and asked him if he could help her daughter. She was giving a presentation at school the next day and wanted to talk about the stars in the Milky Way. But she was getting into a terrible state, because she knew so little about it.

My friend was only too pleased to help out. That same evening, he was sitting in his neighbour's kitchen opposite her daughter. The girl told him that she had once tried to count the stars in the night sky, but kept getting stuck because there were so many. She asked him if he happened to know how many stars there were in the universe.

The astronomer stood up, asked where the store cupboard was, and took out a packet of sugar. He told the girl that it contained around five million grains of sugar. Sixty thousand of these packets – a whole truck full – would amount to 300 billion grains. 'That's how many stars there are in the Milky Way,' he said. 'But the universe is much bigger. We believe that there are at least 100 billion galaxies, the equivalent of 100 million trucks full of sugar grains. If all those trucks were placed bumper to bumper, side by side, they would cover the whole of Europe. The total number of stars in the visible universe, the total number of grains of sugar in all those trucks, is ten trillion. That is a 1 followed by twenty zeros!'

A few days later, the astronomer saw his neighbour and asked her how her daughter's presentation had gone.

'Well, the day after you came around, she was completely upset,' she told him. 'I kept her home from school. After listening to your story, she had spent hours in bed worrying about it. At breakfast, she said she hadn't slept a wink and, to be honest, that's how she looked. All those stars had made her head spin.'

Her daughter had given her presentation a week later. On seal cubs.

WHAT HAD happened to his neighbour's daughter is understandable. It is impossible to imagine such large numbers. You will already lose sleep if you try and see the entire contents of the universe as sugar in trucks. It is even more mind-boggling to try and envisage the empty space between all those stars.

It is more sensible to try and cover astronomical distances in steps, as though you are climbing a staircase. With each new step, you think of things that are a little bigger. You start with a relatively small distance, for example, from the Earth to the Sun, and then compare that to the distance from the Sun to the other planets. Then you move on to the next step and compare the distances in the solar system to those to the stars. In that way, step by step, you get some idea of the dimensions of the solar system, the Milky Way and the rest of the universe. Very quickly, you realize why Bruno's claim that there are also planets orbiting other stars took so long to prove: the stars are simply too far away.

ONE SPRING day in Haarlem, I am standing in front of my class looking out at the historic church of St Bavo. Fifteen second- and third-years are staring at me with a glazed look in their eyes. I have just told them that today we are going to build a scale model of the solar system. With food. I start with the central body in the solar system, the Sun.

'This red cabbage is the Sun,' I say, laying the big round vegetable on the desk in front of me.

We start with the planet that is closest to the Sun: Mercury, the baby of the solar system. I pick a grain of sugar out of a bag and place it next to the cabbage.

Then come Venus and the Earth, two peppercorns. Then another grain of sugar for Mars. For the two largest planets, Jupiter and Saturn, I use grapes. Lastly, I place two raisins on the desk, for Uranus and Neptune. Pluto lost its status as a planet in 2006 and was demoted to a dwarf planet, otherwise I would have added a grain of sand.

There it is on my desk, the solar system. The grains of sugar that represent Mars and Mercury can hardly be distinguished from each other. The planets and the Sun lie close together, as they are always portrayed in books. It looks very cosy, all those little balls in a line, like a family portrait. We often talk about 'our' solar system, as though we have an intimate relationship with our neighbouring planets. But that is a misleading image. If I were to position the planets in proportion to their distance from each other and the Sun, the desk would soon be too small. That feeling of 'our' solar system would soon make way for one of a great emptiness.

To depict this emptiness correctly, I ask my pupils to move the planets so that they are at proportionate distances from the Sun. I tell them that the Sun and the planets on my desk are exactly ten billion times smaller than they are in reality. I remind them that ten billion is a 1 with ten zeros. The Sun, which in reality has a diameter of 1.4 million kilometres, will then be 14 centimetres across – the size of a red cabbage. On the same scale, Mercury, with a diameter of almost 5,000 kilometres, is half a millimetre, the same as a grain of sugar. To make the model as accurate as possible, not only the planets but the distances between them have to be scaled down by the same factor, by scrapping ten zeros from each number. Mercury is 58 million kilometres, or 58,000,000,000 metres, from the Sun. That means that the grain of sugar would have to be 5.8 metres from the red cabbage. Would that fit into the class?

The pupils set to work. A boy places the Sun in the middle of the room. One of the girls says that is no good, because Mercury will never fit into the room. According to her, it's better to put the Sun in one corner. The measuring tape I've brought with me isn't long enough, so the pupils have to measure out the distances in stages. After some messing around, disagreement, moving tables and repeated measurements, the grain of sugar finally ends up on a chair, exactly 5.8 metres from the red cabbage.

Venus doesn't fit into the room: the planet is 108 million kilometres from the Sun, 10.8 metres in our scale model. A second-year boy knocks on the door of the adjoining classroom, gives the bewildered teacher the peppercorn that represents Venus, and asks him to put it on his desk. The other peppercorn, the Earth, is placed on the other side of that room, 15 metres from the cabbage. Mars, as a grain of sugar, is 23 metres from the cabbage, under the willow tree in the school playground. To place the remaining planets, we find a map of Haarlem and take a walk through the town centre. The grape-planets Jupiter and Saturn end up in shop windows. For Uranus, we have to walk a little further, leaving the raisin on a post 287 metres from the school. Then we get to the big church of St Bavo, whose tower can be seen from all over the town. There, exactly 450 metres from our classroom, we place the last raisin: Neptune.

Back in the classroom, we look out of the window. We can see the pointed spire of the big church, hundreds of metres away. I hold the red cabbage in the air, so that it is in an imaginary line with the spire. A raisin near that spire is rotating around a cabbage in our classroom. The Earth and the other planets, as peppercorns, grains of sugar and grapes, are moving on their own orbits somewhere in between. Other than that, there is nothing, except for a few dwarf planets, comets and other pieces of space debris, all of which are much smaller than the smallest sugar crystal in our model. The solar system is empty and lonely.

IT IS essential to know about these distances to understand the universe and the worlds within it, because they make that reality three-dimensional. But how do we know these almost impossible to comprehend distances? How were we ever able to measure the scale of the solar system and the universe?

In classical antiquity, estimating cosmic distances was still the exclusive domain of philosophers. One of them was Archimedes, who most people know for his famous cry of 'Eureka!' on discovering that the level of water in his bath rose when he displaced it with his body. The story goes that he jumped straight out of the bath and ran naked through the streets, shouting with joy.

In his book *The Sand Reckoner*, the ancient Greek streaker attempted to estimate the number of grains of sand that would fit into the universe. Archimedes, who lived in the third century BC, based his calculations on the model of the cosmos devised by one of his contemporaries, the philosopher Aristarchus, who came from the Greek island of Samos. This heliocentric model was at loggerheads with the established belief in the Earth as the centre of the universe. Because all celestial bodies seem to move around us, while we on Earth remain still, many people saw the Earth as the natural centre of the universe. Aristarchus claimed, however, that the movement of celestial bodies could also mean that the Earth moved around the Sun. In his model, the Earth was on different sides of the Sun in summer and winter. That meant that our position in relation to the stars would change a little every day. The position of a star would therefore shift slightly in the sky with the seasons, a phenomenon known as the parallax.

The parallax is the *apparent* change in position of an object when seen from different perspectives. The closer the object, or the greater the distance between the observers, the greater this change seems to be. You can test this yourself by holding your arm stretched out in front of you and looking at your index finger with one eye at a time, keeping the other one closed. You will see that the position of your finger jumps in relation to the background. This apparent change of position – the parallax – increases the closer you hold your finger to your face.

No one in classical antiquity, however, had actually seen a star display a seasonal parallax. That meant there were two possibilities: either the stars were so far away that the parallax was immeasurably small, or the universe was geocentric after all, with the Earth at its centre. If the Earth was immobile, there would be no stellar parallax to be seen.

In *The Sand Reckoner*, Archimedes did not make it clear which of the two possibilities he felt was right. He simply said that the universe must be immensely large. He calculated that 10^{64} grains of sand would fit into the universe (a 1 with 64 zeros), a figure based solely on assumptions about distances in the solar system. The distance from the Earth to the Sun was many times greater than the diameter of the Earth; Archimedes then assumed – incorrectly – that the same ratio applied to the distance from the Earth to the stars and that between the Earth and the Sun. Archimedes' calculation was actually more a demonstration of his mathematical prowess than a measurement. At that time, in the third century BC, there were hardly any instruments or methods accurate enough to measure the parallax.

Archimedes' universe stopped at the immobile 'ceiling' of stars that surrounded the solar system like a kind of disco ball. In the sixteenth century, Giordano Bruno saw it differently. He assumed that everything in the cosmos moved, including the stars. The stars only appeared not to move because they were so incredibly far away. This insight was, as we now know, both correct and revolutionary. And yet, for many centuries, the cosmos was seen not as a spatial whole with three-dimensional depth, but as a still life, an immobile canvas above our heads. It was simply inconceivable that something could be so far away that its movement could not be seen.

Measurements of distance, which would put a stop to this discussion for good, were not yet possible, precisely because the stellar parallax is so immeasurably small. Both Ptolemy's geocentric and Copernicus' heliocentric models saw the night sky as a fixed ceiling that was simply 'far away'. Kepler brought the planets closer to home. His model of the solar system correctly predicted the relative distances between the oval orbits of the planets. It enabled him to calculate that Mars is around one and a half times further away from the Sun than the Earth is. He also reasoned that the stars, because of their apparent immobility, are a lot further from the Earth than the other planets in the solar system. But the *absolute* distances, in kilometres, were still unknown. Kepler's patron, Tycho Brahe, had tried to measure the parallax of Mars – the minimal shift that the red planet makes in the sky as a result of the Earth orbiting the Sun. But that was beyond even this genius of precision.

The parallax of a star, in the constellation Cygnus, was not successfully measured until the nineteenth century, enabling its distance from the Earth to be determined. Before then, there was simply no measuring equipment good enough. As with Bruno's ideas on exoplanets, the theory preceded the observations.

In the meantime, however, there had been a number of remarkable pioneers. In the seventeenth century, two estimates of cosmic distances were made, both of which proved later to have been astoundingly accurate. They may have been lucky guesses, but the methods used to obtain them were as original as they were perilous. One estimated distances within the solar system, the other the distance to the stars. In both cases, a key figure in the development of modern science played a special role. That man was a Dutch clockmaker.

CHRISTIAAN HUYGENS is without doubt the greatest scholar that the Netherlands has ever produced. He was exceptionally versatile: he invented the pendulum clock, was the first to apply a mathematical formula to physics and devised laws of motion on which Newton based his theory of gravity. He also developed the wave theory of light, discovered Titan, Saturn's largest moon, and was the first to describe the nature of the planet's famous rings. Each one a groundbreaking discovery, and in widely varying disciplines.

Christiaan's father was a politician, as was his grandfather. Both had been advisers to the Prince of Orange. His mother died at an early age, leaving his father to bring up the children. Because of their background, the two eldest sons, Christiaan and his brother Constantijn – one year his senior – were destined to a life in politics. With all the European conflicts of the seventeenth century, there was certainly enough to keep them occupied, and the brothers had plenty of opportunity to develop their talents.

Constantijn, his father's favourite, was reserved and loved drawing and music. From a young age, he was introduced to his father's political friends, in preparation for his future as a diplomat. According to his biographer, C. D. Andriesse, Christiaan looked up to his older brother. He also quickly proved to have a great talent for mathematics. It is very likely that he inherited this gift from his mother Suzanna, who his father – Constantijn senior – referred to as 'Sterre' (star) in his poems. Christiaan learned about mathematics by browsing through the large bookcase in the stately family home on the Plein, the large square in the centre of The Hague. He also received lessons from a formidable mathematics teacher by the name of Jan Stampioen. By the age of eleven, Christiaan was already putting together simple machines in his hobby room and asking questions that his father was unable to answer ('the stinking breath of a dog has got into you', his father is reported to have said after checking an exercise by Christiaan that was too difficult for him to understand).

The two brothers went to Leiden to study law. Their overly concerned father gave them a long list of rules to make sure they organized their lives properly. It told them, from hour to hour, what prayers they had to recite, what lessons they had to prepare, when it was time to play music and, not to be forgotten, when it was time to sleep. On this list, Andriesse notes drily, 'it makes one wonder whether they actually stuck to it'. The first letter that sixteen-year-old Christiaan sent home says enough. In short, the message was 'studies going fine, send more money'. Every parent with children at university will recognize concise reports like this.

No matter how much his father would have liked to see him become a successful lawyer or politician, at Leiden Christiaan made a definitive choice for mathematics. His old teacher Stampioen had given him an extensive list of books to get him started on his deeper study of the natural sciences. At the bottom of the list, Stampioen advised his pupil to ensure he put what he read into practice, rather than 'always sitting endlessly with your nose in a book'. Christiaan took this advice to heart.

In his contact with people, Christiaan was timid and reserved, but in his scientific work he was daring and self-confident. At the age of eighteen, he refuted the argument underlying one of Galileo's geometric propositions. The famous French mathematician Marin Mersenne wrote to Christiaan's father, 'if he carries on like this, he will surpass Archimedes himself', causing the proud father to nickname his young brainbox 'my little Archimedes'.

While they were studying, the brothers corresponded regularly with each other and their friends about their escapades in Leiden and later in Breda, the city where Christiaan continued his studies. The older Constantijn in particular seemed to have enjoyed himself enormously. The reserved artist had grown in to an unruly student. Using a secret language, he wrote about nocturnal assignations with girls. More than once, he used the word *vogelen* ('birding') as a codeword for having sex (or in some cases, not, as in: 'She refused to *vogelen*'). His brother Christiaan was far less licentious. He remained a bachelor for his entire life, or rather, was married to science.

It is easy to see the two brothers as two stereotypically opposing personalities: Constantijn the sensitive musician, the womanizer, the amiable politician. And Christiaan the model student, the shy inventor, the reserved genius. But, despite his many other diversions, Constantijn was certainly interested in his brother's activities. In 1654, after being brusquely rejected for a position at the court of Amalia, the widow of Prince Frederik Hendrik of Orange, he was unemployed and had little prospect of finding a new job. So he helped Christiaan, who had by now become interested in telescopes, to grind lenses. Together, they crafted a 3-metre long telescope that enabled Christiaan to discover the striking flat ring of gas around Saturn, and Titan, Saturn's largest moon.[7] The brothers also came up with a new innovation, the tubeless telescope. By linking two lenses that were nearly 4 metres apart in separate casings, they were able to enlarge images much more than had been possible before. Eventually, the brothers built telescopes up to 75 metres long, spectacular constructions that were exhibited in several places around Europe.

Christiaan's inventions were good not only for science, but for his purse. Elaborating on his mathematical equations, he found a way of suspending a pendulum so that it moved back and forth in constant periods. He immediately saw the potential for using his discovery in pendulum clocks, and patented his design.

Huygens's 'tail clock' quickly became popular. In the seventeenth century, the Netherlands' 'Golden Age', the fight for world domination was fought on the high seas, and clocks were essential for navigation. A ship's latitude can be directly pinpointed from the position of the Sun at its highest point, but anyone sailing from East to West passes through different time zones. To find your way then, besides the positions of the stars, you need an accurate reading of the time. It was therefore logical for any expedition to the East or the West to have a tail clock on board.

IN 1666 Huygens was appointed to the French Académie Royale des Sciences. The contacts he made gave him the opportunity to determine whether his tail clock was indeed suitable for finding ships' positions at sea. One of his acquaintances was the Italian astronomer Giovanni Domenico Cassini (Jean-Dominique to his French friends), who became the first director of the Paris observatory in 1671. Cassini specialized in measuring distances, which brought him a number of notable assignments. Around 1700, by order of Louis xiv, he measured the size of France. The country proved (typically enough) to be much smaller than the French had hitherto assumed. There is an apocryphal story that this earned Cassini a good-natured reprimand from the Sun King. 'You have deprived me of more of my Kingdom', Louis is alleged to have said, 'than I have gained in all of my wars.' Even after Jean-Dominique's death, the Cassini family continued to make maps for the French court for many centuries.

Because of Cassini's affinity with defining positions and measuring distances, Huygens's tail clocks were taken along on one of the first purely scientific maritime expeditions in history. Such voyages were typically prone to disaster. At the time, disease, shipwrecks, desertion and encounters with cannibals were less the exception than the rule. In 1669, Jean Richer, a promising young scientist at the Académie, was selected for a scientific voyage across the Atlantic Ocean. On board were a number of

Huygens's clocks to determine the ship's longitude accurately. But it was not a success: the clocks stopped working at the first sign of substantial waves and the ship's crew were unable to get them working again. In a letter to a colleague, Huygens complained about Richer, saying that 'a drop of oil' would have got the mechanism going again. He blamed the failure of the pendulum clocks on the nonchalance of the crew rather than on the clocks themselves, though he did admit later that the tail clocks did indeed not work optimally at sea.

In 1672 Cassini sent Richer on a new expedition, this time to Cayenne in French Guiana. One of the expedition's main aims was to conduct the measurement that had long been astronomers' Holy Grail: the Mars parallax. The plan was for Richer to measure the position of Mars from Cayenne, while Cassini would do the same from Paris at exactly the same moment. Assuming that Mars is closer to the Earth than the stars, the planet would show a different position in respect of the surrounding stars in the two measurements. In terms of the experiment with your extended index finger, you would see it against a different background with your right eye (from Paris) than with your left (from Cayenne). This difference of position – the parallax – increases with the distance between the observers. That is why Richer was sent to the jungle. Since France and South America are so far apart, it was assumed that the parallax would be large enough to measure.

The great challenge now was to ensure that the measurements were made at exactly the same time. That called for a clock. Huygens's tail clocks had proven unsuitable for sea voyages. So they had to find another way of measuring the time simultaneously thousands of kilometres apart. The heavens themselves presented a solution. On 22 August 1672, there was to be a solar eclipse, which would be visible in both Europe and South America at more or less the same moment. Richer and the observers back in France could therefore compare the time they measured Mars's position with the time they saw the eclipse. In the same way, Cassini and Richer used the positions of Jupiter's four moons as a reference point in time. This method proved a success.

SCIENTISTS TRAVELLING to far-off places sparked the imagination of those back at home. A drawing by court painter Sébastien Leclerc shows Richer in his observatory in Cayenne. We see the astronomer dressed in a robe and taking measurements on a globe. The light of the late afternoon sun through the open window falls on the French furniture. There is a sextant to determine the position of celestial bodies. A tail clock on the wall betrays another scientific aim of the expedition: the pendulum proved to move more slowly back and forth here than in Paris. This phenomenon was used to show that gravity is weaker there than in Europe, a possible indication that the Earth is not a perfect sphere.

The drawing is very probably a romanticized portrayal of reality. Leclerc had clearly never been in South America. Richer's accommodation was very primitive: you would be more likely to find a hammock there than a chaise longue. The scientist's calm appearance is also very improbable, as he spent a large part of this time in the tropics in bed, shivering with a fever.

The measurements of Mars were not the only scientific objective of the expeditions. Richer also made a small number of biological observations, but his studies of the indigenous flora and fauna are less than impressive, to say the least. He noted that dolphins are warm-blooded, while turtles are not. He investigated whether the opening on the back of a pecari, a species of tropical boar, was related to its breathing (this proved not to be the case). He determined that crocodiles can live without food for many months and took one back to France with him to prove it. The unfortunate reptile died during the voyage.

Dead crocodile or not, Richer was given a hero's welcome when he arrived home in 1673. Bernard le Bovier de Fontenelle, a prominent writer of popular science, described his return with a heightened sense of drama in the annals of the Académie: 'People awaited Richer's return as though waiting for a judge to pronounce his ruling on the difficult questions that divided astronomers. You could say that astronomy was in a state of suspense when Richer came back from Cayenne.'

When Richer's and Cassini's simultaneous measurements of Mars were compared, the planet indeed appeared to have shifted in relation to the background stars by some 15 arc seconds. The measurement of the parallax meant that the angles and the short side of the Paris–Cayenne–Mars triangle were known, enabling Richer to calculate the distance from Mars to Earth. And with that measurement, he could use Kepler's laws to calculate the orbits of the planets and the distance from the Earth to the Sun. He came up with a distance of 140 million kilometres, an unprecedentedly accurate figure.[8] Richer might not have been much of a biologist but, like Tycho Brahe, he was a brilliant astronomer. The distance we use today, 150 million kilometres, was not measured until a century later, in 1769, when the British captain James Cook observed a rare phenomenon from the Pacific island of Tahiti: the planet Venus passed across the face of the Sun. By comparing Cook's observation with observations of the transit made in Europe, the parallax could be used to determine the absolute distance from the Earth to the Sun.

Back to Richer. When Fontenelle gets to the Mars parallax in his report, he starts waxing lyrical, calling it the '*Grande Affaire*' of Richer's expedition:

There is nothing more pleasing or wonderful about mathematical truths than their extreme fecundity. The 15 arc seconds of the Mars parallax can hardly be distinguished by eye or instrument, but they give us the enormous expanse of the solar system, [hundreds of millions of kilometres] . . . A margin of uncertainty of three arc seconds, no more than a hair's breadth, represents a measurement error of hundreds of thousands of kilometres in the distance to Mars. These distances are formidable, or rather, our [previous] conceptions of space were too small. Why do we call something that exceeds our imagination formidable? . . . Since when do our ideas set the limits?

What Fontenelle describes is a significant change of mentality. Where distances and proportions in the cosmos used to be a matter of guesswork, it was now possible to prove that the universe is immeasurably large. Robust measurements and hard figures convinced even the most hard-headed sceptics, and the implications of the measurements by Richer and Cassini found their way, through Fontenelle, to the general public. The immeasurable solar system had finally been measured. That changed people's view of the world. Where they had once been at the centre of the cosmos, the Earth and its inhabitants were now, in one fell swoop, small and insignificant. Voltaire, Enlightenment philosopher *par excellence*, put this idea into words half a century later, in 1747. In the book of the same title, his character Zadig considers the night sky and comes up with a profound metaphor: 'He thus imagined men such as they are, in effect: so many insects devouring one another on a little atom of mud.'

Christiaan Huygens, too, had heard about the results of Richer's second voyage in 1673. As a well-known figure in Parisian social circles, he announced during an afternoon lecture at the mansion of a Duchess the news of 'numbers, curves, touching planes and triangles that he had received from Cayenne'. Although he was more likely referring to the pendulum period of his tail clock, we can conclude that Huygens must have been aware of the Mars measurements.

THESE CELESTIAL discoveries were made against the background
of terrestrial discord. In 1672, while Richer was setting up his
measuring instruments in Cayenne, Holland was experiencing
an *annus horribilis*. Four countries, including England and France,
had declared war on the Republic of the Netherlands.

Two days before the solar eclipse that enabled the measurements of Mars to be conducted simultaneously, the brothers
and statesmen Johan and Cornelis de Witt were lynched by an
enraged mob in The Hague. In the words of a popular saying
from the time, 'the people were beyond reason, the government
beyond hope and the country beyond saving.' The people of
Holland had more on their minds than astronomical observations
in the tropics.

At the end of the decade, it had become impossible for
Huygens to remain in Paris, because of the political situation.
The globalization of science had been put on hold by worldly
affairs. Towards the end of the Golden Age, Huygens was back
living in the Netherlands, first with his father on the Plein in the
centre of The Hague, and later in a country house in Voorburg,
just outside the city. He still had contact with other scholars, but
disagreed with them on many issues. He was not impressed, for
example, with Newton's theory of gravity. Perhaps he became –
as is often the case with scholars in the autumn of their careers
– something of an old grouch. He missed his brother Constantijn,
who was by then working at the English court. He wrote to him
that he was preparing a 'philosophical essay', which he would
dedicate to him. In the book, entitled *Cosmotheoros*, he measured
the distance to the stars and travelled them in his mind.

Galileo had discovered the wonders of the solar system with
his telescope. Kepler had calculated its dimensions using the
accurate measurements of his old employer, Tycho Brahe. The
measurements of Cassini and Richer more or less confirmed

those dimensions. The question that now occupied Huygens was, how far away are the stars? It was already known that the parallax of the stars was immeasurably small, suggesting that they were incredibly distant. It was necessary to be creative and find another way to measure that distance. And Huygens discovered how to do that. 'Little Archimedes' lived up to his pet name by estimating – just like the Greek sand reckoner – the distance to the stars. But this time by measurement, rather than by pure guesswork. He devised a method that was brilliant in its simplicity and precision; it did not require a single measuring instrument. It was to be one of his final experiments.

The reasoning was simple. Imagine that all stars were the same as the Sun: the same size, equally massive, and equally hot, and all with the same light intensity. The only factor that thus determines the *apparent* (as observed by us) differences in their luminosity is their distance from the Earth and thus the size of the dot we see in the night sky. A bright star is therefore closer to the Earth than a fainter one.

Huygens wanted to compare the brightness (and hence the apparent size) of two stars to calculate their difference in distance. One was the Sun and the other was Sirius, also known as the Dog Star. Sirius is the brightest star in the firmament. As the Sun appears much brighter and bigger than Sirius, Huygens reasoned that it must be closer to the Earth. Exactly how much closer could be calculated by measuring their difference in brightness.

But how did Huygens go about determining exactly how much brighter the disc of the Sun was than the dot of Sirius? That was no easy task, especially in an age without digital cameras or electronic light metres. Huygens used his memory to measure the difference in brightness. He would look at the Sun during the day through a very small hole. The hole had to be so small that the sunlight that passed through it matched the brightness of Sirius at night. The ratio between the size of the Sun and that of the small hole would be the same as their relative distances from the Earth. After all, an object that is twice as far away, is twice as small.

Our brains have no absolute sense of light sensitivity, just as we generally have no absolute sense of hearing, feeling or taste. Few people can remember how the musical note A sounds, feel how warm a shower is, or taste how many grains of sugar there are in their tea. Nevertheless, Huygens tried to remember how bright Sirius was. He studied the star at night from his back garden and imprinted its image on his memory. Perhaps he compared it with other sources of light, like the Moon, a candle or another star. The following day, with that memory fresh in his mind, he would commit the greatest astronomical sin possible and look directly at the Sun. Galileo had gone blind earlier for the same reason. But Huygens had taken preventive measures.

Back in his garden, which was now bathed in sunlight, he had covered his head on all sides, 'so that no Light might come near my eye to hinder my Observation'. Then he looked at the Sun through a telescope tube. At the other end of the tube was a thin plate with a small hole, through which the sunlight reached his eye. But no matter how small he made the hole, the small point of sunlight that came through it was still much too bright. Only when he also placed a condensing lens against the hole did he obtain the desired result. From under his dark covers, the point of light was exactly as bright as the Dog Star that he had seen the previous night.

He calculated that the small hole and the lens together reduced the image by a factor of 27,664. That meant that the disc of the Sun was 27,664 times bigger than the dot of light that was Sirius. Consequently, Sirius was 27,664 times further away from the Earth than the Sun was. Thanks to Cassini and Richer, the distance from the Sun to the Earth was known to be 140 million kilometres. On this basis, Huygens concluded that Sirius was 4 trillion – 4,000,000,000,000 – kilometres away, a distance that can rightly be called astronomic.

The accuracy of this measurement is as miraculous as the simplicity of the experiment. The actual distance to Sirius from the Earth, which we have now measured using the parallax method, is 80 trillion kilometres, some twenty times further than Huygens's estimate. The difference is not the result of an incorrect measurement, but an erroneous assumption: that all stars are the same. Sirius is in reality 25 times brighter than the Sun and therefore seems to be five times closer – almost equal to the degree of Huygens's error. In short, you can measure the size of the universe from your back garden with a tube, a few blankets and a small lens. And a sharp set of eyes.

THIS UNIQUE experiment takes us back to the classroom in Haarlem, where my edible model showed just how formidable distances in the solar system really are. The Sun is a red cabbage, the Earth a peppercorn 15 metres away, Neptune a raisin 450 metres further away. Space would become even more lonely, if that is possible, if we worked out how far from the Sun the closest star would be. That is not Sirius, but Alpha Centauri, the brightest star (indicated with the first letter of the Greek alphabet) in the constellation Centaurus. Alpha Centauri is a little over 40 trillion kilometres from Earth – 41,306,000,000,000 to be precise. To make it easier to work with such enormous distances, we use light years – the distance it takes for light to travel in one year. One light year is a little less than 10 trillion kilometres, so that Alpha Centauri is just over four light years away. On our model of the solar system, which fits into the centre of Haarlem, we would not encounter our second red cabbage until we were 4,000 kilometres away, say somewhere near Tehran, the capital of Iran.[9]

It is perhaps comforting to know that Alpha Centauri has a companion. In our scale model, there is another star only 200 metres further away, Alpha Centauri B. I am not very familiar with the centre of Tehran, but let us say that star A is a red

cabbage at the mosque and the slightly smaller star B is an artichoke on a vegetable stall in the bazaar. The two stars orbit each
other, just as the planets orbit the Sun. In fact, the Alpha Centauri
group is a trinary system, but the third star, Proxima Centauri,
is so close to the other two – only 200 kilometres in our scale
model (2 trillion kilometres in reality) – that their gravitational
pull causes it to circle them in a small orbit.[10]

In 2012 a planet was discovered orbiting Alpha Centauri B.
It is about the same size as the Earth but is much closer to its
mother star. In our model, this planet would be a peppercorn
in a crate next to the artichoke. It is unlikely that life is possible on the planet, as it is simply too hot for complex molecules
to develop. It is, however, quite possible that there are undiscovered, more habitable planets orbiting the stars in the Alpha
Centauri system. This begs the question of whether we could
ever go there, or whether any aliens living there could visit us.

TO GIVE us an impression of how long it would take to reach
Alpha Centauri, we will take a trip into space Hollywood-style. In
the film *Gravity*, Sandra Bullock and George Clooney are repair-
ing the famous American Hubble Space Telescope when they
are hit by a storm of space debris. In one of the most blood-
curdling moments at the start of the film, Bullock is separated
from the shuttle and floats through space at high speed. It is
every astronaut's greatest fear.

In this situation, she would in reality continue to orbit the
Earth like a satellite. At a height of 550 kilometres, she would
complete an orbit every hour and a half, at a speed of 27,000
kilometres per hour in relation to the Earth. Let us now ignore
the laws of nature for a moment and imagine that she moves
away from the Earth at the same speed in a straight line. George
Clooney doesn't save her and Sandra floats on and on out of
the solar system. After about four months (I fear that by now

she would no longer be conscious), she would reach Mars. It would then take her another two years to reach Jupiter. Saturn and Uranus would take yet another two and a half and six years. Eventually, eighteen years after her unfortunate separation from the shuttle, she would reach Neptune. After that, she would float for another ten years or so through the Kuiper Belt, a ring of small and larger lumps of rock, including Pluto. In the Oort Cloud, which extends a quarter of the way to Alpha Centauri, she would occasionally encounter a comet that belongs to the solar system. Then there would be nothing for quite a while. Sandra's remains would not reach Alpha Centauri B and its planet until 160,000 years later.

It seems very unlikely, given these enormous distances, that we will be able to travel to other planets in the near future. But we can always dream about it. And that is exactly what Christiaan Huygens did at the end of his life, after making his measurements of Sirius. These cosmic distances gave him ideas that he shared with his brother: he speculated on the possibility of extraterrestrial life.

three

AN INQUISITIVE MIND AND DEFECTIVE SIGHT

C*OSMOTHEOROS, OR The Celestial Worlds Discover'd: Conjectures Concerning the Inhabitants, Plants and Productions of the Worlds in the Planets* was the world's first science-fiction bestseller. Even the Russian tsar Peter the Great read the book during his stay in the Netherlands. He found it very entertaining, and commissioned a translation into Russian as soon as he returned home. The conservative religious printer called it a 'satanic perfidy' by a 'delirious author', but the tsar was insistent and the translation went ahead. The book fuelled the imagination of the people of Europe, who showed a great interest in all things extraterrestrial.

Cosmotheoros was the 'philosophical essay' that Christiaan Huygens had written at the end of his life and dedicated to his brother Constantijn. At that time, Constantijn was secretary to stadholder-king William III in London. He was at Christiaan's bedside when he died in 1695, after being ill for a few months. The essay was published three years later in book form. The title, *The Celestial Worlds Discover'd*, perfectly encompasses its content. The book contains ingenious measurements (such as the distance to Sirius) and sharp analyses of the solar system. It describes an imaginary journey to the planets and speculates on the 'plants and productions' to be found on them. Huygens then caps it all by speculating about the possibility of extraterrestrial life. He wastes no time in getting to the point:

A Man that is of Copernicus's Opinion, that this Earth of ours is a Planet, carry'd round and enlighten'd by the Sun, like the rest of them, cannot but sometimes have a fancy, that it's not improbable that the rest of the Planets have their Dress and Furniture, nay and their Inhabitants too as well as this Earth of ours.

As Christiaan writes, this idea was not new to the brothers:

This has often been our talk, I remember, good Brother, over a large Telescope, when we have been viewing those Bodies, a study that your continual business and absence have interrupted for this many years.

Christiaan is remembering the good old days, when the brothers were in their twenties, grinding lenses together, building telescopes and making discoveries in the firmament. Just like every other stargazer, they clearly philosophized and speculated on the question: are we alone in the universe? In the following chapters of *Cosmotheoros*, Christiaan takes his brother on a journey through space, starting with the planets of our own solar system. He describes how the Earth must look from the Moon: 'The Earth to them must seem much larger than the Moon doth to us, as being in Diameter above four times bigger.'

Galileo had seen mountains on the Moon through his telescope. The planets could therefore also have similar landscapes. Huygens hypothesizes that, in theory, there could be living beings on all the planets. He applies the scientific principle that, until proved otherwise, the possibility cannot be discounted. On the Sun, life as we know it is improbable because of the great heat. 'Such Bodies as ours could not live one moment in such a Furnace,' he writes, adding immediately: 'We must make new sorts of Animals then, such as we have no Idea or Likeness of among us, such as we can neither imagine nor conceive.'

That is true, of course: life on Earth is not necessarily a blueprint for life elsewhere in the universe. Huygens writes that water, a basic ingredient for life on Earth, may not play any biological role at all in other parts of the universe, saying 'Perhaps their Plants and Animals may have another sort of Nourishment there.'

Elsewhere, he notes that inhabitants of other planets do not necessarily have to resemble people:

> For 'tis a very ridiculous opinion . . . that it is impossible a rational Soul should dwell in any other shape than ours . . . Yet methinks this fancy has such a rule upon my mind, that I cannot without horror and impatience suffer any other figure for the habitation of a reasonable Soul.

Huygens continues that the image of a 'Creature like a Man' but with 'great round sawcer Eyes' arouses the utmost aversion, 'altho at the same time I can give no account of my Dislike'. His observation that our prejudices may give us a limited notion of alien life is significant and way ahead of its time.

After having demonstrated the enormous scale of the universe by measuring the distance to Sirius, Huygens stresses that our planetary system cannot be unique. Why should other stars not also have 'attendant' planets around them? Huygens fantasizes about the nature of life on these planets, singing their praises in poetic terms:

> They must have their Plants and Animals, nay and their rational ones too, and those as great Admirers, and as diligent Observers of the Heavens as our selves . . . What a wonderful and amazing Scheme have we here of the magnificent Vastness of the Universe! So many Suns, so many Earths, and every one of them stock'd with so many Herbs, Trees and Animals, and adorn'd with so many Seas and Mountains!

His sense of wonder becomes even greater when he realizes how many stars there are and thus how many 'adorned' planets there could be. We recognize here Giordano Bruno's ideas on 'infinite worlds', and Huygens indeed referred to Bruno, using the rather prosaic name 'Jordaan Bruin'.

The multiplicity of worlds was thus no new idea, but *Cosmotheoros* was revolutionary for several reasons. Firstly, Huygens's hypothesis on extraterrestrial life is not simply plucked out of the air. He uses his measurement of the distance to Sirius to illustrate how large the universe must be. He also notes that more powerful telescopes would lead to the discovery of new, more distant stars. Building on this basis, he then asks his readers to think beyond Earthbound parameters and imagine alternative life forms. Nor does he exclude the possibility that some of these alien life forms may be intelligent. Sometimes his speculations seem ludicrous. The 'Planetarians' he says, are skilled in not only astronomy and writing, but geometry, arithmetic and the 'mechanical Arts'. Furthermore, he claimed that they have compasses and sailing ships (there is, after all, wind on the planets), golden jewellery (there is also gold on the planets) and pendulum clocks (there is gravity on other planets). They would, he supposed, 'enjoy not only the Profit, but the Pleasures arising from such a Society: such as Conversation, Amours, Jesting, and Sights'. We should not take such musings too seriously. The point Huygens wanted to make was that there is no reason at all to assume that 'Planetarians' – and their multifarious activities – do not exist.

Cosmotheoros became a bestseller in Dutch, German, French, English, Swedish and Russian. A hundred years after it was first published, it was still being reprinted. In more recent times, however, the book is somewhat downplayed in accounts of Huygens's life. Some biographers find all that rambling on about Planetarians undignified for such a great scientist. In 1977, for example, Jan and Annie Romein wrote:

It was certainly not his most important work: it contained no new discoveries and the scientific value of this undeniably well-considered and astute argument can justifiably be called into doubt . . . This short book, both naive and intelligent . . . raises the question: is this really the work of the man who taught the seventeenth century to think scientifically? . . . It is easy to answer this question by saying that such signs of old age are common . . . It is then up to succeeding generations to decide whether this is the product of the wisdom of years or a natural decline of mental capacities.

Such an assessment is somewhat unjustified and even insulting. When we read *Cosmotheoros*, we can hear Christiaan Huygens chuckling between the lines. Admittedly, the assertions he makes are sometimes a little 'bold', to use his own words, but the crux of the matter is clear: the universe is immensely large. According to the laws of logic and the large numbers involved, it could easily be filled with planets, and they could be teeming with life. Prove it otherwise, Huygens challenges us in return.

HUYGENS WAS not the first serious astronomer to try his hand at science fiction. A century earlier, Johannes Kepler had written *Somnium* (The Dream), a novel about the inhabitants of the Moon, which was also published posthumously. Besides *Somnium*, Huygens was also inspired by *Entretiens sur la pluralité des mondes* (Conversations of the Plurality of Worlds), which was published in 1686 and would sell even better than *Cosmotheoros*.

Entretiens was written by Bernard le Bovier de Fontenelle, whom we encountered earlier as the chronicler of the French Académie Royale des Sciences. Like Huygens, Fontenelle had his choice of career imposed upon him by birth. In his case, it was the law. And like Christiaan, while studying, he also became fascinated by the natural sciences. He devoured the works of the ancient Greek mathematician Euclid and the Renaissance philosopher René Descartes. Unlike Huygens, however, he did not possess a great talent for mathematics; Fontenelle primarily had a passion for writing. At first, he wrote plays, which were all dismal failures. According to the playwright Jean Racine, Fontenelle's greatest achievement in theatre was that, during a performance of his first play *Aspar*, the audience responded for the first time with a chorus of catcalls and whistles. Until then, Parisian theatre-goers would express their displeasure only by yawning and throwing rotten apples.

Fontenelle decided on a radical career change. Six years after the disastrous premiere, he published the book that was to make his name. *Entretiens sur la pluralité des mondes* is a series of dialogues between a well-spoken philosopher – undoubtedly a self-portrait of the author – and the beautiful Marquise de la Mésangère. Their conversations take place on the Marquise's estate, during nocturnal walks beneath the starry sky: an excellent setting for a French romantic novel. But the narrator is exceptionally discreet about any potentially romantic developments, admitting coyly: 'suffice to say that I had the pleasure of being her only visitor.'

We read the book now as one big cliché. The delicate lady hangs on every word uttered by the all-knowing philosopher, who reveals to her the wonders of the universe. It is an accurate portrayal of the times: in the eighteenth century, women were kept completely out of public life. Many men saw the natural sciences, not yet taken seriously as a discipline, as a suitable way for them to pass the time. It is therefore not surprising that the first women to appear in scientific records are mathematicians or astronomers.

Fontenelle cleverly took advantage of this situation by making one of the main characters in his book a woman. Consequently, *Entretiens* became especially popular among wealthy ladies. It is a very accessible and lightly speculative text. In the foreword,

Fontenelle promises a combination of philosophy and amusement, in which he combines truth with fiction. Like Bruno, his starting point is the supposition that stars, like the Sun, have planets orbiting them. He also speculates on the possibility and nature of extraterrestrial life. He says that aliens do not necessarily need to be intelligent, making a comparison that sounds decidedly racist:

> When adventurers explore unknown countries, the inhabitants they find are scarcely human; they are animals in the shape of men, even in that respect sometimes imperfect; but almost devoid of human reason.

The colonial spirit of the times even resounds in futuristic fantasies about space travel.

The book sold like hot cakes. It was reprinted more than twenty times in Fontenelle's lifetime alone. By way of comparison, Copernicus' *De Revolutionibus*, which introduced the heliocentric model and on which all science fiction is arguably based, has only been reprinted five times in four centuries. The Church never devoted attention to the controversial opinions expressed in *Entretiens*. Fontenelle was a little more tactful than Bruno and Galileo. He reached a respectable old age by never stepping on the wrong toes and, in his own words, by eating large quantities of strawberries. He died a month before his hundredth birthday.

A quote from *Entretiens* neatly summarizes the search for extraterrestrial life before the twentieth century:

> All philosophy is founded on two things; an inquisitive mind, and defective sight; for if your eyes could discern everything to perfection you would easily perceive whether each star is a sun, giving light to a number of worlds; on the other hand, had you less curiosity, you would hardly take the trouble to inform yourself about the matter, and consequently remain in equal

ignorance; but the difficulty consists in our wanting to become acquainted with more than we can see: besides, it is out of our power to understand much of what is within the reach of our sight, because objects appear to us very different from what they are. Thus philosophers pass their lives in disbelieving what they see, and endeavouring to conjecture what is concealed from them; such a state of mind is not very enviable.

This brings us back to the handicap that also afflicted Bruno: what stops us from finding exoplanets is not a lack of curiosity but the inability to see far enough. That is why no one looked for them for many centuries, though many people speculated on what lay beyond the bounds of what we cannot see. It may sound crazy to us now, but in the eighteenth century the idea of life on other planets in the solar system was widely accepted, as was the conviction that there were planets orbiting other stars, which could also be home to forms of life. James Ferguson, a much-read astronomer, wrote in 1756:

> Since the fixed stars are prodigious spheres of fire, like our sun ... it is reasonable to conclude that they are made for the same purposes that the sun is – each to bestow light, heat, and vegetation, on a certain number of inhabited planets.

The most striking example of a renowned scientist who believed in aliens was Sir William Herschel. The discoverer of Uranus and the most celebrated British astronomer – three moons and planetary craters, one planetoid, countless observatories, a space telescope, schools and streets throughout Europe, a village in Canada and a pub in Slough are named after him – was convinced that there was extraterrestrial life on the Sun. In 1801 he proposed that the Sun had a rocky surface covered by a blanket of clouds that was very hot on the outside. The sunspots that he observed as black dots on the surface must be holes in the cloud blanket. Under that hot atmosphere, the climate would undoubtedly provide a very pleasant environment for life to exist. Herschel concluded that 'we have reason to look upon the sun as a most magnificent habitable globe.'

It was not discovered until a century later that the Sun's heat comes from its interior and that it consists entirely of hot gas. The temperature on the surface is much too hot for solid ground to form, not to mention organic life. The sunspots are indeed somewhat cooler areas on the surface, as Herschel suspected, but they are caused by the Sun's magnetic field and not by holes in the clouds. The possibility of life on the Sun has since been ruled out, but Herschel could not have known that.

The popularity of alien life persisted unabated. William Herschel's son John, like his father (and his aunt Caroline Herschel), was a famous astronomer. He discovered moons around Saturn and Uranus and introduced a new form of measuring astronomical time, which is still used today. In 1833, he left Portsmouth on a ship bound for the Cape of Good Hope. There, he aimed his telescope at the southern night sky, which had been charted in the sixteenth century by Dutch seafarers De Houtman and Keyser, but was still relatively unexplored by telescope three hundred years later.

John Herschel stayed on the Cape for five years. When he came home, he was welcomed as a hero. He was showered with titles and guest professorships and received a firm handshake from the Queen. The study of the southern hemisphere suddenly doubled the field of play for astronomers. But the most striking news that Herschel had sent back from South Africa was the discovery of living beings on the Moon.

In August 1835, the *New York Sun* – a small local newspaper – published a travel account by a Dr Andrew Grant, who had accompanied Herschel as a scientific assistant. The article began with a triumphant list of his master's discoveries. Herschel had formulated a new theory on comets, discovered planets in other solar systems and had, in his spare time, 'solved or corrected nearly every leading problem of mathematical astronomy'.

Grant then described Herschel's study of the Moon using the revolutionary new telescope he had invented. The first thing Herschel had seen were blue bison, enormous herds of them lumbering across the rock-strewn lunar plains. Then he discovered two-legged beavers that lived in huts, from which smoke curled into the air, and described more than a hundred strange kinds of flora and fauna in great detail. The jewel in the crown was *Vespertilio-homo*, a hairy, winged humanoid creature. Through the eyepiece of his telescope, Herschel had seen groups of these beings conducting animated conversations – there was no doubt at all that this was an intelligent life form. Unfortunately, after his discovery, the absent-minded professor had left his telescope out too long in the Sun and the lens had become unusable. The world would have to wait for further information on life on the Moon.

The news exploded like a bombshell. In cafés and salons in New York, there was talk of nothing else. The story was reprinted in local and national newspapers and soon spread to Europe. Herschel was a popular figure and his expedition had already attracted a great deal of publicity, so the public lapped up this new story. For centuries, there had been speculation about extra-terrestrial life, but now it had been discovered – and by a famed astronomer at that.

In scientific circles, however, there was considerable scepticism about Herschel's story. It was easy to refute it: even with the most powerful telescopes of the time, it would be impossible to distinguish vegetation on the Moon, not to mention two-legged beavers. The dry voice of reason was lost, however, in the tumult of cheerful public speculation about life on the Moon.

Very soon, however, the real author of the article was revealed: satirical writer Richard Locke had concocted the whole story. The practical joke was a successful gamble on the gullibility of the public. The Great Moon Hoax was the first and best viral prank in the history of science. Herschel himself allegedly saw the funny side of the affair, though he did formally distance himself from the story later, after being overloaded with questions from people fascinated by his 'discoveries'.

Even after the man in the street was aware of the truth, many people continued to believe in the men on the Moon. Paintings and drawings of the bat-like creatures still adorned magazines and snuff boxes. Edgar Allan Poe revisited Locke's prank in a slightly less dramatic version, in the form of 'The Balloon Hoax': a fictional, but authentic-sounding report of a manned hot-air balloon that had crossed the Atlantic Ocean in three days.

Even the Dutch writer and Protestant minister Nicolaas Beets heard about the 'Moonmen'. It was immediately clear to him that every letter of these 'absurd fairy-tales' was a fabrication. But that did not mean that there were not edifying lessons to be learned from them. Shortly after the Hoax, in 1836, Beets published a science fiction poem entitled *Mannekens in de Maan* (Men in the Moon). Like Huygens, Beets had no illusions about what the aliens looked like:

They have been spotted. – Whether they were handsome boys,
That, my dear girl! is another matter
I seem to recall unkempt hair.
And beards, hanging down to their bellies.

But what concerned him the most was that the Moonmen grew up wild:

As you can see, they are still barbarians!
Development and civilization come slowly.
Through the telescope, there was nothing to be seen,
of school buildings, theatres or churches.

And, of course, here too the colonial spirit resounds:

And everyone can see what has to happen here
It has never yet occurred on Earth,
That a newly found tribe is left in peace.
And that must not take place here, either.

The poem closes with a look ahead to the Apollo project:

We must – in the interests of the 'hordes'
That live there – go to the Moon forthwith!
I have no doubt at all that humankind will go there,
And likely sooner than many would dream of.

BOTH SERIOUS astronomers and popular authors saw extraterrestrial life as the norm. So there was certainly no lack of inquisitive minds. But what about defective sight? Were people already looking for planets around other stars? Giordano Bruno had already identified the source of the problem: the stars were simply too far away to discover much smaller planets orbiting them. And that is still the case. Even with the best twenty-first-century telescopes, it is only possible to capture images of a handful of exoplanets. Exoplanets can best be found indirectly, by looking closely at the stars they orbit. And, in those times, a lot of people examined the stars very closely.

Newton's theory of gravity showed that the laws of nature applied not only on Earth but in space. And yet astronomy continued to be largely distinct from physics until deep into the nineteenth century. It was mainly astrometry, precisely measuring the positions of stars and planets. Astrophysics did not emerge until later. The first methods of discovering planets around other stars were therefore devised by astrometrists.

Astrometry is a laborious discipline, with long nights spent watching the sky, peering at measuring instruments and twiddling knobs. It is a profession for hermits, bachelors and obsessive collectors. Publications from the time give a good impression of this astrometric subculture. It was clearly something of a sport to observe, like Tycho Brahe, as many stars and planets as possible and as often as possible, and to record their positions. Some astronomers, many of them amateurs, built up catalogues of tens of thousands of stars, which they monitored for long periods of time.

Astrometrists jumped on everything that moved in the cosmos. In the first instance, of course, that was the parallax, the seasonal difference in the position of the stars in the sky. Because of the motion of the Earth around the Sun, we see the stars from a slightly different angle in the winter than in the summer. The stellar parallax was first measured successfully in 1838 by a fanatical German astrometrist, former office clerk Friedrich Wilhelm Bessel. He measured the parallax of the star 61 Cygni (a star in the constellation Cygnus, the Swan)[11] as three-tenths of an arc second (0.3 arc seconds, as small as a peanut on Tower Bridge viewed from Big Ben). That put the star twelve light years away from Earth.

Other moving objects that were often studied were comets, binary stars and planetary moons. Binary stars were a favourite of the British Captain William Jacob, an army engineer and head astronomer at the observatory in Madras, now Chennai, India. The observatory was set up by the British East India Company. Astronomy and seafaring were traditionally intimately related, as navigation depended on the positions of the stars. Captain Jacob used his many hours on night duty in India to make astronomical observations, especially of binary stars.

Binary stars are so close together that their gravity ensures that they visibly orbit each other. Some are too far away from us to be distinguished separately, but relatively closer binary star systems like Alpha Centauri can be seen through a telescope as two dots of light and their movements can be followed separately. Captain Jacob followed the movements of the two stars in another close binary system, 70 Ophiuchus, in the constellation Ophiuchus, the Serpent Bearer. Besides the expected circular dance of the two stars around each other, one of the stars also exhibited an additional movement. It was an extra 'wobble' with a much shorter period than that of the binary orbit. Jacob suspected that it was caused by the gravity of a third object, invisible to his telescope. He proposed that this dark object could be a planet. In an article for the Royal Astronomical Society, he wrote:

> There is, then, some positive evidence in favour of the existence of a planetary body in connection with this system, enough for us to pronounce it highly probable, and certainly good ground for watching the pair closely, to procure, if possible, still stronger evidence.

With these words, Captain Jacob went down in history in 1855 as the first person to claim, albeit cautiously, to have discovered an exoplanet. His suggestion to watch the system closely was taken to heart. The wobble was, however, so small that it could hardly be distinguished from the inaccuracies that occur in any measurement of position caused, for example, by refracted light in the atmosphere or shifts in the position of the telescope. The subjective opinion of the observer also plays a significant role in how such uncertainties of measurement are interpreted. In the period up to 1900, various publications appeared confirming or refuting the existence of a planet around 70 Ophiuchus.

One of these publications, which came out in 1895, caught the attention of the wider public. It was not the conclusion of the article – the proof of a 'dark companion', that is, a planet, orbiting 70 Ophiuchus – that caused the greatest stir, but its author. Thomas Jefferson See was the son of a large family of farmers, who had been given the opportunity to study at the University of Missouri. He developed a keen interest in astronomy, and the university observatory became his second home. He spent night after night there gazing at comets, planets and stars. He was chosen as the best student of the year and gave the Valedictory Address at the graduation ceremony. It would not be his last speech. See liked to be the centre of attention.

Photographs offer a reasonably good impression of his character: a man in a three-piece suit with a prominent moustache, posing next to a telescope, staring at a point in the sky with an intense and theatrical expression. Science alone was not enough; See wanted his discoveries to be made known to the public at large. Besides countless articles in academic journals, he also wrote regularly for popular scientific magazines. Although this is in itself a very noble and important goal, See's desire for sensation often got the better of his scientific integrity.

Two years after producing his 'proof' of a 'dark companion' around 70 Ophiuchus, See published an article in the popular magazine *Atlantic* with the somewhat pretentious title 'Recent Discoveries Respecting the Origin of the Universe'. In an expansive explanation of the origins of the solar system, See referred to practically all renowned scholars of the previous two centuries, ending with himself as their natural successor. He summed up his own work on binary star systems and noted his own recent remarkable discoveries, saying 'If they should turn out dark bodies in fact, shining only by the reflected light of the stars around which they revolve, we should have the first case of planets – dark bodies – noticed among the fixed stars.' Despite the somewhat cautious wording, See is clearly convinced of the revolutionary nature of his discovery.

Two years later, See's claim was blown out of the water. One of his former students, Forest Ray Moulton, calculated that the system portrayed by See could not exist. The hypothetical planet was so massive that its gravity would distort the orbit of the binary star. Such a system of three celestial bodies would be very unstable. See was enraged. Three days after Moulton's article appeared in the *Astronomical Journal*, he sent a letter to the same publication. He went straight for the jugular, ripping Moulton's 'interesting application' to shreds. The journal only published the first paragraph, claiming that the rest was 'not related to Mr. Moulton's article'. This was followed by a sharp reprimand from the editors:

> The present is as fitting an opportunity as any to observe that heretofore Dr. See has been permitted, in the presentation of his views in this journal, the widest latitude that even a forced interpretation of the rules of catholicity would allow; but hereafter that he must not be surprised if these rules, whether as to soundness, pertinency, discreetness or propriety, are construed within what may appear to him unduly restricted limits.

In other words, they had had enough of See's theatricals. That was the last time the *Astronomical Journal* would publish an article by his hand.

Later, See would make his way into the press again with theories on the origins of stars, earthquakes, the ether and an alternative form of gravity. He was an exceptionally hard critic of Einstein, calling his theory of relativity a 'crazy vagary'. Every modern-day scientist receives an email every now and again from someone who claims to have developed a new 'theory of everything' that solves all the problems in the cosmos. The usual response is to click on the delete button. The difference with See was that he had a group of devoted followers who hung on to his every word whenever he presented a new breakthrough. Yet, time and time again, he failed to come up with any mathematical foundation for his claims. Consequently, perhaps rightly, history has marked him as a charlatan.

So what exactly makes See a charlatan, while Huygens and Herschel – despite their speculations about extraterrestrial life

– are still seen as the greatest scientists of their time? There are two important differences. Firstly, Huygens and Herschel produced a large quantity of real results that helped move science forward, while See achieved very little in terms of science. Secondly, Huygens's and Herschel's ideas on extraterrestrial life are clearly speculative, while See presented the existence of the 'dark companion' as the inevitable consequence of his measurements. That was not the case, as he had insufficient measurements to provide incontrovertible evidence. His emotional response to Moulton's well-constructed counterargument did nothing to enhance his credibility.

We still do not know whether there is indeed a planet orbiting 70 Ophiuchus. A search using modern techniques in 2006 yielded no conclusive evidence, but the existence of a 'dark companion' close to one of the binary stars cannot be excluded. If there is one, it will be much smaller than See imagined, because it does not significantly distort the orbit of the binary star.

THE STORY of 70 Ophiuchus contributed to the search for exoplanets no longer being taken very seriously in scientific circles. Wealthy businessman-turned-astronomer Percival Lowell, one of See's former employers, also made a blunder that did little to improve the situation. On the basis of very unclear images, Lowell claimed with a great deal of pomposity that lines he had observed on the surface of Mars could be nothing other than canals dug by an intelligent civilization. The response in the press was unequivocal. 'MARS INHABITED, SAYS PROFESSOR LOWELL', shouted the headline in the *New York Times* of 30 August 1907. Astronomers at Harvard ignore him, the article wrote, but 'the people of this country support Prof. Lowell in his Martian campaign.'

This report heralded the start of the twentieth century. Planet hunters were the darlings of the public, but were dismissed as pariahs by their scientific colleagues. Anything to do with extraterrestrial life was seen as belonging to the realms of 'little green men'.[12] The 'canals' on Mars proved to be merely an optical illusion, but the public's imagination had been sufficiently fired up. A golden age dawned for science fiction. *The War of the Worlds*, H. G. Wells's book about an invasion of Earth by Martians, was made into an immensely popular radio play in the 1930s. There are still cock-and-bull stories going the rounds of an outbreak of mass panic in 1938 because people took the radio reports of bloodthirsty space creatures too seriously. Later, of course, came *Star Trek*, a series about space voyages to exoplanets that still has a large following of almost religious fans, called 'trekkies'. Films like *Alien*, E.T. and *Star Wars* have become modern legends. In short, the popularity of speculating on life on other planets has not diminished since Huygens's time.

ASTRONOMY TOOK a different course in the twentieth century. Astrometry was the old science, and astrophysics had the future. Quantum mechanics was developed in theory and by experiment in the early years of the century. The direct cause of this was that observations on the nature of light could not be fully explained by the existing laws of nature. The same applied to the light from the stars. Better instruments made it possible to discover countless phenomena in light that provided direct information on the structure of stars. Astronomers were able to look further into the universe and, in the 1920s, Edwin Hubble discovered not only that it was much larger than had previously been believed, but that it was expanding. A whole new world opened up, literally. We no longer measured only the positions of stars, but also the natural laws that determined their past and their future.

Amid the tumult of these great discoveries, hunting for new planets disappeared from the astronomical agenda – inasmuch as it had ever been on it. This temporary exile was illustrated by the story of one of the last pure astrometrists, Peter van de Kamp.

Born in the old Hanseatic city of Kampen in 1901, Van de Kamp – a gifted pianist – chose to go to university rather than the music academy. After completing his studies, he moved to America for good, even declining an offer to become director of the astronomical institute at the University of Utrecht in 1937. He became a professor at and later director of Swarthmore College, a small but leading college in an idyllic university town in the state of Pennsylvania, where he would remain for more than forty years. Towards the end of the Second World War, he returned to Europe for a short while where, together with a group of fellow scientists of different nationalities, he took part in the Alsos Mission to find out how far the Germans had progressed in developing the atom bomb. The scientists ventured into enemy-held territory ahead of the advancing liberation forces to interrogate their German colleagues as quickly as possible.

Van de Kamp was a tall and charming man, with wide-ranging interests beyond astronomy. At Swarthmore he lifted the ban on public musical performances, a remnant of the college's strong Quaker roots. He became the conductor of the college's symphony orchestra, accompanied silent films on the piano and even once played the viola in a quartet with the renowned amateur violinist Albert Einstein.

The science that Van de Kamp practised was traditional astrometry *par excellence*. He was very passionate about measuring the positions of stars, which was extremely painstaking work. He would record the distance between stars endlessly on photographic plates. Like Jacob and See, he discovered wobbles in the movements of stars, which he attributed to 'dark companions'. His measurements were of a very high quality and all the companions he found have since been confirmed. Most are a kind of intermediate category of body between giant planets and stars, which would later be given the name 'brown dwarfs'. But there was one dark companion that was to become his arch-enemy.

From the time of his appointment at Swarthmore in 1938 to deep into the 1970s, Van de Kamp monitored Barnard's Star – the second closest star to the Sun, after Alpha Centauri, at some six light years' distance. Any wobbles in its movement, caused by a dark companion, could therefore be clearly observed. In 1963, after tens of thousands of measurements over several decades, Van de Kamp reported the existence of a wobble in the movement of Barnard's Star. His measurements proved that there must be a celestial body with a mass of around one and half times greater than that of Jupiter orbiting the star. Van de Kamp concluded that a companion of such mass could not be a star but had to be a planet. The discovery made him one of the most famed astronomers in the world. The *New York Times* wrote hopefully: 'The new finding adds support to the conviction of astronomers that a great many solar systems exist, some of them possibly supporting life.'

Van de Kamp chose a successor to continue the measurements of Barnard's Star, the German Wulff Heintz. When taking the measurements, however, Heintz started to have his doubts. He noticed that the photographic plates were not accurate enough to make measurement of the motion of the star possible. That proved to apply also to Van de Kamp's older observations, which Heintz retrieved from the archives. From 1976, he published a series of articles calling the discovery of the planet into question. He proposed that large-scale maintenance of the telescope in 1949 had resulted in the image shifting – a movement that Van de Kamp interpreted as the movement of the star.

Van de Kamp responded with indignation. He saw the attack, and especially by the successor he had chosen himself, as a stab in the back. He received a lot of support, leading Heintz to recall later: 'I was denounced among his friends – including top administrators – as a nasty character and probably mentally disturbed.' Nevertheless, Heintz's claim that Barnard's Star displayed no observable wobble was confirmed by dozens of independent studies. Ironically enough, Barnard's Star is so stable that it is used today as a reference point to test whether a telescope is firmly fixed. If observers see a wobble, they know that it is the telescope that is wobbling and not the star. But Van de Kamp remained embittered by the affair for the rest of his life and continued to maintain until his death that Barnard's planet existed. *Eppur si muove.*

It was a regrettable episode. No one doubted Van de Kamp's qualities as an astrometrist. John Gaustad, a fellow professor at Swarthmore, noted that it is 'important to keep in mind, though, that, in terms of the rest of Van de Kamp's career, he did very important, accurate work, and that the field of astronomy is richer for it.' The dark planet around Barnard's Star proved to be a blind spot. Staring at one object for his whole life meant that even an honest and painstaking scientist like Van de Kamp lost his objectivity. Heintz put a lot at risk by denouncing his predecessor, but his measurements left him no choice. The scientific method knows no mercy.

FOR THREE centuries, our inquisitiveness was aroused. Huygens's wandering stars, Fontenelle's many worlds, Herschel's Sun beings and Beets's Moonmen, Lowell's canals on Mars and finally the exoplanets of *Star Trek* – extraterrestrial life was alive and kicking in popular culture.

But defective sight spoiled the party. After the Van de Kamp episode, planet hunters became extinct as a profession. Heintz and his successors showed that photography alone was not capable of finding planets around other stars. Yet scientists continued to hope that they would find life elsewhere in the universe. It had proved too difficult to find other planets, but perhaps that was not even necessary. Maybe E.T. would call us himself. And that meant that we had to make every effort to ensure we could receive the signal loud and clear when he did.

four

THE ORDER OF THE DOLPHIN

T HE FILM *Independence Day* was for me the absolute high point of the year 1996, packed as it was with action and scary extra-terrestrial beings, an attack on the White House and Will Smith saving the world with a cigar in the corner of his mouth. For weeks on end, everyone in my first-year class at secondary school talked of nothing else.

The film starts with a shot of the American flag on the Moon. In the background, a lonely trumpet sounds. The camera pans downwards to show a commemorative plaque bearing the text 'Here men from the planet Earth first set foot upon the Moon, July 1969 AD. We came in peace for all mankind.' We hear the crackling voice of Neil Armstrong speak the same words. We see his footsteps in the lunar dust.

Then the ground starts to tremble. A shadow passes slowly over the plaque and the flag. The music builds up ominously. The camera pans slowly upwards. First, we see the Earth. Then, accompanied by the sound of bombastic wind instruments, a gigantic flat spaceship comes into view, heading straight for our planet.

In the next scene, we see a row of satellite dishes in the desert of New Mexico. It is the SETI Institute. SETI stands for the 'Search for Extraterrestrial Intelligence'. The Institute filters radio signals from space, in the hope of finding signs of intelligent life. The search has been going on for many years, but nothing has ever been found. In the film, we see a lonely technician at the Institute doing the night shift. He is bored stiff and is practising his putting on a grass mat in his office. In the background, we hear Michael Stipe of R.E.M. singing 'It's the end of the world as we know it.'

Then suddenly, a red light starts to flash on one of the receivers and we hear a loud beep. The analyst can't believe his eyes. He stumbles over his own feet as he rushes to the telephone and calls his supervisor. Rudely awakened, his boss says 'If this isn't an insanely beautiful woman, I'm hanging up.'

'Shut up and listen!' the technician says, breathlessly. He holds the telephone up to a speaker, which is emitting a radio signal that sounds like Morse code. The supervisor jumps out of bed, banging his head. He mumbles something about yet another bump and that the signal probably comes from a Russian spy satellite. Still in his dressing gown, he totters into the control room, stumbling over a couple of golf balls on the floor. The technician, wild with excitement, says 'This is the real thing. A radio signal from another world!'

SETH SHOSTAK, an astronomer at the SETI Institute, was delighted at the time that the makers of the film had even heard of the Institute's existence. But he was a little indignant about how his place of work had been portrayed. He went to his boss to complain about the errors in the blockbuster movie. First of all, the Institute is not in the desert but in a high-tech district of Silicon Valley. The film suggested that the staff consisted of little more than one or two technicians, while the Institute actually employs around a hundred people. 'And that stuff about the golf,' Shostak said. 'It makes it look as though we sit around here twiddling our thumbs!' His boss didn't reply, just calmly brought something out from behind his desk. A golf putter.

Shostak is a comedian, the Robin Williams of astronomy, with straight, grey hair, a friendly face, dark eyebrows and one corner of his mouth permanently curled upwards, as though he is about to tell a joke. Which he mostly does, in a sing-song voice that is pleasant to listen to. A radio voice. Everything he says is witty, slightly mocking and optimistic. The latter in particular is a very useful character trait if you spend your whole life waiting for a telephone call from E.T.

When Shostak was young, 'like all kids' he was interested in two things: dinosaurs and aliens. 'We are programmed to be fascinated by things that may be a threat to us,' he tells me. 'That's why so many programmes on Animal Planet are about crocodiles and sharks, rather than *eekhoorns*.' He uses the Dutch word for squirrel, a leftover from his years working on the Westerbork radio telescope in the Netherlands. He also used to run a company that made computerized animated films.

The screensaver on Shostak's computer flashes up photographs from his career. Radio telescope dishes, satellite images, Shostak with Will Smith and on Larry King's talk show. Seth Shostak is the public face of SETI. He writes books and does one-man shows on the search for extraterrestrial life. He also presents a weekly radio show from the SETI Institute, in which he interviews a wide variety guests on how they see science. 'I never speak longer than seven minutes,' he says. 'Otherwise it gets boring. We've been sitting here for half an hour already. Aren't you bored? Hmm. Maybe it's an American thing.'

On the day of my visit, the subject of the radio show happens to be related to SETI. The guest is Denise Herzing, a biologist who has spent thirty years studying how dolphins communicate. Shostak calls 'Dolphin' an example of an extraterrestrial language that is spoken here on Earth. Like whales and apes, dolphins are important in studying the development of intelligent life. Intelligence appears able to evolve in different ways. Herzing believes that it is very possible that, after humanity disappears, it will be replaced by a different form of intelligent life. According to Shostak, racoons are in with a good chance.

The show also features another frequent guest, Frank Drake. Drake is one of the founders of SETI. He likes to talk about his hobbies – cultivating orchids and lapidary (cutting and polishing gemstones) – but even more so about the quest that has taken up more than fifty years of his life, so far without success. It is a quest that started with eleven dolphins.

IT IS difficult to identify a specific moment at which SETI started. That is partly because it is an abstract concept. It is not a telescope or an organization – the SETI Institute was not set up until later. SETI is more like a movement. Its full name, the Search for Extraterrestrial Intelligence, refers to the search in its broadest sense. Anyone can conduct a SETI experiment. That is to say, anyone with a radio telescope.

The seed for SETI was sown on a spring day in 1959, during a philosophical conversation between two physicists at Cornell University in the state of New York. The Italian Giuseppe Cocconi had stepped into the office of his American colleague Philip Morrison. During the Second World War, Morrison had been part of the Manhattan Project, the group of scientists who developed the atom bomb. After the bombing of Hiroshima and Nagasaki, he became a fervent opponent of the spread of nuclear weapons.

By then, the Space Age was well underway. The Russians had sent the first satellite into space, and the race for the Moon had begun. Science fiction fuelled speculation about what we would find in outer space. And that also influenced the two physicists. Cocconi wondered whether the radiation that is released during a nuclear explosion – and could be produced in a particle accelerator at their university – could also be used in theory to communicate in space. They quickly came to the conclusion that it would present too many practical difficulties. And then the most interesting question arose: imagine that extraterrestrial beings wanted to contact us. What would be the best form of communication for them to use?

Radio waves seemed the most suitable, as they can penetrate the atmosphere of a planet like the Earth without any problem. A radio station on Earth only uses one frequency with a very narrow band width. If you turn the tuning knob on your radio a little to the left or right, the channel you are listening to gives way to interference. If we pick such a single-frequency signal from a star, we've struck lucky, as it means that it must have been broadcast by radio equipment – the star itself cannot emit a signal with such a narrow bandwidth.

Cocconi and Morrison published their thoughts in the authoritative scientific journal *Nature*, in an article entitled 'Searching for Interstellar Communications'. This article is still seen as SETI's historical manifesto. The authors claimed that it was possible that, just as on Earth, an intelligent civilization had developed on another planet. And it might even be possible that this civilization was older and more advanced than ours. If these aliens also felt the need to communicate with life forms elsewhere in the universe, it was very likely that they would also use radio waves. That is a lot of 'possibles', but the authors emphasized that none of them could be rejected on scientific grounds. So they called for an extensive and targeted search for radio signals in space.

Cocconi and Morrison did not know that, at the same moment, 600 kilometres away, a young radio astronomer had come up with the same idea, and was even putting it into practice. Years before, Frank Drake, who was working at Green Bank, West Virginia, had picked up an unexpected radio signal from the direction of a star. There proved to be a simple explanation, but Drake could not shake off the thought that extraterrestrial life could be using radio signals. He conducted the first SETI experiment ever, Project Ozma (named after the witch from the *Wizard of Oz*), using the large radio telescope at his institute. Green Bank lies in a thinly populated and green part of the U.S., where a large proportion of the inhabitants have hoofs. It is an ideal spot for radio telescopes, because there are few man-made transmissions in the air to cause interference. Frank Drake pointed the radio telescope at two stars, Epsilon Eridani and Tau Ceti, for a total of 150 hours. Both stars are less than ten light years from the Earth. Drake had calculated that a radio signal from that distance would in theory be detectable by the 26-metre-diameter dish at Green Bank.

For the whole of those 150 hours, the stars maintained radio silence. There was a signal, but it unmistakably came from the star itself. The experiment produced no signs of extraterrestrial life, but it did attract a great deal of attention.[13] A year later, Drake received a request from the National Academy of Sciences to organize a conference on the search for extraterrestrial life. He was allowed to compile the list of participants himself. He reminisced on this during a lecture in 2010, saying 'I convened a meeting of all the people in the world who I knew were thinking about extraterrestrial intelligent life. All twelve of them. And they all came.'

The conference was held at Green Bank in November 1961, right next to the radio dish Drake had used for his experiment. The meeting was kept somewhat secret as, at that time, the subject was not considered worthy of serious scientific attention. There was no official announcement; nor were the minutes published afterwards. Two days before the first session – the first guests had already arrived at Green Bank – there was not even an agenda. Drake put one together at the last moment. It was to become the most famous meeting agenda in the history of astronomy, and was even given a nickname, the Drake Equation. Fifty years later, every astronomer around the world can recite it in their sleep.

The aim of the conference was to determine how worthwhile it was to search for extraterrestrial life by trying to pick up radio signals. Drake considered how to estimate the number of intelligent civilizations in the Milky Way – our cosmic backyard – that might want to establish contact with us. He wrote down all the factors that might be significant and put them together to form a mathematical equation. If the result was significantly greater than zero, the SETI experiments would not be a waste of time.

How many intelligent civilizations that can communicate by radio are there in the Milky Way? The Drake Equation divides this complex question into manageable secondary questions that bring us, step-by-step, closer to the answer. You start with the total number of stars in the Milky Way.[14] Then you estimate how many of these stars have planets, and the average number of habitable planets per star – that is, planets with a solid surface and a favourable temperature – and on what percentage of these planets life will evolve. The list of questions stops here if you are interested only in whether extraterrestrial life exists in any form. But, because SETI experiments are only focused on signals from intelligent life forms capable of communicating, Drake went a step further and asked in how many of these cases we are talking about intelligent life. How many of these intelligent life forms develop radio equipment and send messages into space, intentionally or by accident? Lastly, the timing also has to be right: how many of these intelligent beings exist at this moment? That entails estimating the average lifespan of a civilization.

When Drake was finished, he had seven symbols in a row, all of which represented one of the above estimates. It would be a matter of trial and error, as none of the figures were known. For some of them, the estimates still vary widely. But the equation gave Drake and his guests at Green Bank a main thread for their discussions. A separate session was devoted to each of the seven symbols, and there were one or more experts present when each estimate was made.

FIRST OF all, there was Otto Struve, the first director of the radio observatory at Green Bank since 1952. He had given Drake permission to conduct the Ozma experiment. Struve was the fifth generation of a German-Russian dynasty of astronomers. His great-grandfather had worked with Joseph von Fraunhofer, the maker of the telescope through which a stellar parallax was first observed. During the First World War, the Struves had fled to the Crimea to escape the Bolsheviks. After the war, at the age of 23, Otto ended up in a Turkish refugee camp. He worked as a lumberjack and lived in a small tent with five others. A letter of distress, sent to an uncle who had already died, ended up after a rather unlikely detour in the hands of the director of the Yerkes Observatory in Wisconsin, 150 kilometres from Chicago. The director immediately offered Struve a job. He arrived at Ellis Island, New York, in 1921 as one of the many European immigrants who would never leave the u.s. again. Otto Struve was to become one of the most proficient and influential astronomers of the twentieth century. He was the director of several observatories, including Yerkes. He was known as an excellent teacher, popular with his students, but also as a demanding boss. He was the first to arrive in the morning and the last to leave at night, and he also kept close track of his staff's working hours. The students he took on included a large number of immigrants who, like Struve, had come to try their luck in the New World. Two of them would later win the Nobel Prize: the Indian astrophysicist Subrahmanyan Chandrasekhar, whose theoretical work formed the basis for research into black holes, and the German-Canadian physical chemist Gerhard Herzberg, an expert on molecules.

Struve believed strongly in the existence of exoplanets and by no means excluded the possibility of extraterrestrial life. This belief was not the product of a lively imagination, but was solidly founded in science. Struve was the first person to observe that stars turn on their axes. And he discovered that they all rotate too slowly. According to the prevailing theory – which was later confirmed – stars were formed from rotating gas clouds, which collapsed under their own gravity. Struve proposed that, if the rotation – the formal term is 'angular momentum' – of an imploded cloud were eventually to be transferred to the stars, they would end up spinning very rapidly. His observations showed, however, that this was not the case. The angular momentum must therefore have gone somewhere else; Struve concluded that it must have been transferred to a planetary system. Our own solar system was a good example of this: the mass and orbital period of Jupiter shows that the giant planet accounts for 99 per cent of the total impulse moment of the solar system. The Sun rotates slowly around its axis, just like the stars that Struve observed. For him, this was an indication that the stars also had planets.[15]

On the basis of this assumption, Struve wrote an article in 1952, only one and a half pages long, in which he proposed two methods for observing planets around other stars. These are now known as the transit and the Doppler methods.[16] Struve thought that it would be almost – if not entirely – impossible to discover an exoplanet with the technology of the time. Half a century later, however, when technology had overtaken Struve's predictive powers, his two methods would be used to discover thousands of planets.

Around the brainstorming table in Green Bank in 1961, together with his former student Su-Shu Huang, Otto Struve estimated that half of all stars have a planetary system, one in five of which house habitable planets. This estimate was rather hit-or-miss, but the presence of planets was at least indirectly based on observation.

THE REST of the invitees around the table in Green Bank were a mixed group. Bernard Oliver, inventor and head of the research department of computer company Hewlett-Packard, had flown in by private plane. Dana Atchley was director of a company whose products included microwaves for the army. He had donated an advanced signal enhancer to Drake's Ozma experiment. Philip Morrison, one of the authors of the SETI manifesto, had also been invited.

There was also another astronomer present. At 27, New Yorker Carl Sagan, a researcher at Berkeley, was the youngest of the group. Drake had earlier exchanged letters with him and knew that he was very interested in astrobiology, the study of the origins and existence of life in the universe. Drake described Sagan as 'dark, brash and brilliant'. Sagan would later become world famous. He was the global face of astronomy and the driving force behind the search for extraterrestrial life, bridging the gap between science and science fiction. Many will remember him as the author of the book *Cosmos* and presenter of the television series of the same name. Sagan always fought to release science from its ivory tower. That earned him the scorn and criticism of many colleagues, who saw him as more concerned with attracting attention than as a serious scientist. Sagan's persistence was, however, of inestimable value in defining and confirming the role of science in society.[17]

Astrobiologist Sagan was joined in Green Bank by two heavyweights of biochemistry, also invited by Drake. Joshua Lederberg already had a Nobel Prize in chemistry to his name, while Melvin Calvin heard during the conference that he had won one, too. Three bottles of champagne, smuggled in secretly, were opened and there was a toast to the good prospects that life would develop on a habitable planet. These experts thought those prospects were very good: after all, life had evolved on Earth as soon as the surface had sufficiently cooled.

The last three estimates – how often the emergence of life led to an intelligent civilization, how many of those civilizations can communicate with radio signals, and how long would such a civilization last? – were perhaps the least reliable made at Green Bank. To answer the first question, how rare intelligent life is, they turned to John Lilly. He was researching communication between dolphins, which he considered an intelligent life form. This meant that on the Earth alone, more than one form of intelligent life had developed. With unwavering optimism, they assumed that intelligent life had developed on all the planets where life had evolved.

There was, however, general agreement that dolphins could not be expected to develop radio communications equipment in the near future. The second question – whether an intelligent life form could communicate using radio – was therefore estimated at 10 to 20 per cent. Again, this was a wild guess, hardly based on fact. The final factor, the average lifespan of a civilization capable of communicating, was the most difficult to estimate. Our own civilization had only been capable of communicating with radio waves for some fifty years. How long would it be before a natural disaster would destroy humanity, or that we would do the job ourselves? The estimate of how long a civilization could communicate was very wide-ranging, from 1,000 to 100 million years.

Putting all these assumptions together, the researchers came up with an estimate for the number of civilizations in the Milky Way currently capable of communication: a minimum of 1,000 and a maximum of 100 million. Millions of the hundred billion stars in the Milky Way could have planets orbiting them with intelligent life forms seeking to make contact with us. That implied that the universe, with more than a hundred billion galaxies, is teeming with intelligent life.

This rather vague estimate was not the most important outcome of the conference at Green Bank. The researchers were very well aware that their calculations were dependent on countless uncertain factors, most of which have still not been determined with any greater accuracy to this day. But Green Bank did produce two important results. Firstly, there was now a framework, a plan of attack to address the issue of life elsewhere in the universe on a step-by-step basis: the Drake Equation. Future research should focus on estimating the values in the equation. Secondly, the provisional conclusion of the guesswork gave reason for optimism. The number was significantly larger than zero, so there was more than enough reason to continue the search.

At the end of the conference, the eleven initiates dubbed themselves 'The Order of the Dolphin', after John Lilly's favourite topic of research. Melvin Calvin sent all the participants a badge of an old Greek coin showing a leaping dolphin. For the rest of their lives, the members of the Order would petition for Project Ozma, Drake's first SETI experiment, to be continued.

IN THE Space Age, the number of sightings of UFOs increased exponentially. This was more likely due to people's growing interest in extraterrestrial life than any actual evidence of aliens being interested in humans. In the 1950s, the term 'little green men' became popular. Every time someone claimed to have encountered aliens, they were little green men. Even when, in 1995, several people in Kentucky reported seeing elongated grey beings in a flying saucer, the newspapers called them little green men (it turned out to have been a meteorite and two rather disgruntled owls). Many years later, during the success of the television series *The X-Files*, there would be another peak in sightings of flying saucers.

In 1967, the SETI experiment took an unexpected turn. Jocelyn Bell, a doctoral student in astronomy, picked up an unusual signal with the radio telescope in Cambridge. The signal repeated itself; every 1.3 seconds, a tick could be heard coming from a dark spot in the sky. A signal of this kind had never before been observed. No known celestial body emitted such a radio signal. As a kind of joke – it was, after all, the Space Age – the source of the signal was called LGM (Little Green Man)-1. Years later, Bell wrote 'here was I trying to get a PhD out of a new technique, and some silly lot of little green men had to choose my aerial and my frequency to communicate with us!'

When Bell and Antony Hewish, her thesis supervisor, published their discoveries in *Nature*, they had already rejected the 'alien civilization' interpretation, as they had also discovered similar sources coming from other locations in the sky. Nevertheless, the suggestion of alien beings seeking to make contact with us was enough to make the discovery world news. Press photographers got a rather bewildered Bell to pose triumphantly with her arms in the air, urging her on and saying 'Look happy dear, you've just made a discovery!'

Eventually, the explanation that Hewish and Bell had already given in their *Nature* article proved correct: LGM-1 was not an extraterrestrial radio transmitter, but a neutron star, a dark and very compact remnant of an exploded massive star, spinning like a kind of lighthouse. These newly discovered celestial objects, which emitted a regularly ticking signal, were given the name 'pulsars'. Because of their extremely high density (they are as massive as the Sun, but as small as Amsterdam), pulsars constitute a challenge to our theories of gravity and matter and they remain the subject of very intensive study to this day. The discovery of pulsars resulted in a Nobel Prize in Physics for Hewish and a number of colleagues. Bell, who had discovered the first pulsar, was not one of them, an omission that many see as the one of the greatest scandals in the history of science. Bell's discovery of LGM-1 continued to fire the popular imagination, and the radio signal appeared on the cover of the debut album of the English band Joy Division.

THE SEARCH for extraterrestrial radio signals was intensified. After several updates of Project Ozma, there was a growing desire for a more radical and large-scale approach. In the 1970s SETI became part of the national research agenda. Besides the U.S., Italy and the Soviet Union are the only countries that have ever invested public money in the search. People are not as generous and optimistic everywhere. When Shostak recently asked an audience in the Netherlands whether they would be prepared to spend the price of one cup of coffee a year on SETI, the collective answer was 'no'.

THE LARGE-SCALE American SETI effort was put in place two years after the Moon landing. In 1971, NASA published a thick report entitled *Project Cyclops*. The authors included inventor Bernard Oliver, involved in SETI since Green Bank, and John Billingham, an English army doctor who had become a medical specialist at NASA. The report stated that there was a growing conviction among scientists that the universe was full of life. On the first page, there was a quotation by Frank Drake that summed up the message of the 250-page volume:

> At this very minute, with almost absolute certainty, radio waves sent forth by other intelligent civilizations are falling on the earth. A telescope can be built that, pointed in the right place, and tuned to the right frequency, could discover these waves. Someday, from somewhere out among the stars, will come the answers to many of the oldest, most important, and most exciting questions mankind has asked.

The authors of the report argued that, if there were to be any chance of success, the current observations would have to be increased manyfold. They believed that there was already evidence of planets outside the solar system. Peter van de Kamp's discovery of a planet around Barnard's Star had not yet been disproved and was presented as an incentive for Project Cyclops. The technical part of the report was written by Bernard Oliver, who had designed a cluster of no fewer than 1,500 radio dishes which would scour the heavens without interruption. But the costs proved a little too steep for the government, and Cyclops was never built. The report did mark, however, the start of NASA's sponsoring of SETI experiments which would continue for more than twenty years.

IN THE 1970s and '80s John Billingham and his colleagues organized regular workshops at Ames, a NASA research centre in Silicon Valley. Astronomers, physicists, mathematicians, biologists and geologists would come together and discuss how they could solve the Drake Equation. They asked the same questions that the Order of the Dolphin had addressed in 1961. One of the participants in the workshops was the young astronomer and computer programmer Jill Tarter. As a child of the Space Age in the 1950s, Tarter had become fascinated by extraterrestrial life while watching science-fiction series like *Flash Gordon*. Walking on the beach with her father in Florida, she would look up at the stars and know for certain that, on a different planet up there somewhere, another girl was walking on the beach with her own father. 'It seemed obvious to me, in the way it can to a child, that we were not alone in the universe,' she recalled years later. She helped her dad fix things in his garage; she liked to take things apart and put them back together again. By the time she was eight years old, she knew what she wanted to be – an engineer.

The dream proved difficult to fulfil. The Tarters did not have enough money to send their daughter to university. Because Jill was a direct descendant of Ezra Cornell, founder of the university that bears his name, she tried to get a scholarship there. She received a letter back saying that scholarships were only available for male descendants. Fortunately, a few days later, the company Proctor & Gamble offered her a scholarship. Consequently, in the early 1960s, she was just about the only woman among the three hundred first-years taking engineering at Cornell. At that time, there was an evening curfew for women at the university: they were not allowed to leave their rooms after ten o'clock in the evening. Tarter therefore had to solve challenging nocturnal problems all alone, while her male fellow students could consult with each other. That may have contributed to her exceptionally unique and original research portfolio.

Because she found the engineering lectures rather dull, Tarter decided to broaden her horizons after completing her studies. In the early 1970s, she enrolled at Berkeley as a PhD student, where one of her specialist subjects was astronomy. She pioneered research into a category of dark objects with less mass than stars – the same objects that Peter van de Kamp had discovered several of. Tarter called these 'almost-stars' brown dwarfs. They are now the subject of close study, partly because they are the big brothers of exoplanets.

SETI discovered Tarter thanks to the Berkeley astronomer Stu Boyer, one of the participants in Billingham's workshops. Boyer had got hold of a computer that he wanted to programme to find extraterrestrial radio signals. It was a commercial micro-computer, a PDP-8/s. Nearly a square metre in size, it just fitted on a desk, making it the first desktop computer. Only a handful of people knew how to operate such a hypermodern machine, and Tarter was one of them. Boyer walked into her office at Berkeley one day and gave her the Cyclops report. She read it from cover to cover, and then again. And again. Many years later, Tarter would recall:

> The plan was a miracle of technical ingenuity. Thanks to the radars developed during World War II, we now suddenly had the tools to answer the greatest question of all. Where we used to have to ask priests and philosophers what we should believe, we could now find aliens ourselves by conducting an experiment. 'Faith' gave way to 'exploration'. I was hooked.

The term 'planet hunter' was born during the workshops at Ames. 'The biologists in the room asked how many planets a star would have on average,' Tarter explained later. 'But we couldn't answer that. Otto Struve had of course made an estimate at Green Bank, but we will only know for sure after we've measured it. At the workshops, we organized the first splinter sessions on how to detect planets.' One of the participants in the sessions was Bill Borucki, a young engineer who had previously worked on the Apollo project. He would later become a key figure in the search for exoplanets.

John Billingham had to account regularly to Congress for the practical usefulness of the search for signals from little green men, whose existence was far from certain. After all, it was funded with taxpayers' money. 'Five hundred years ago, America hadn't been discovered,' he countered. 'For all people knew, it didn't exist.' He argued that the unknown deserved the chance to be discovered.

Billingham's greatest opponent was William Proxmire, a senator famed for his Golden Fleece Awards, which he awarded every month to a project that was, in his eyes, a complete waste of money. In February 1978, the prize went to SETI, which Proxmire called 'one of the biggest, most ridiculous examples of wasteful spending', and proposed to defer it 'for a few million light years'. A few years later, government funding for SETI experiments was discontinued. Senator Proxmire's statement showed, by the way, that he had not done his homework: a light year is a unit of distance, not of time.

Carl Sagan, now a well-known public figure because of his appearances on television, visited Proxmire personally to explain the Drake Equation to him. Sagan played on the senator's fierce anti-war sentiments. A civilization advanced enough to make contact with us, Sagan argued, would certainly possess the technology to build nuclear weapons. If they succeeded in reaching our planet, that would also mean that they had not used their nuclear weapons to destroy themselves. We could therefore learn much from such peace-loving beings. Proxmire was convinced by Sagan's argument. Partly owing to his public support, NASA received funds to conduct SETI experiments until well into the 1990s.

Nevertheless, Billingham and Oliver had by now realized that it would be better for the SETI project to be less dependent on government support in any case. Too many of the funds they received came with a heavy bureaucratic burden. Furthermore, Jill Tarter recalls that SETI was not taken very seriously in the astronomical community. 'When I graduated from Berkeley,' she says, 'people looked down on SETI. Besides searching for radio signals from aliens, the hunt for planets around other stars was also equated with believing in little green men. Astronomy was concerned with the origins of the universe, black holes and other mysterious objects. If you didn't study these, you didn't count. It had been like that since Einstein. Planetary science was the lowest of the low. The question of whether there was life elsewhere in the universe was assigned to the realms of science fiction. Planets just weren't sexy enough for astronomers.'

Within NASA, the hunt for exoplanets and the origins of life was eventually separated from the search for extraterrestrial radio signals. In 1984, Tarter became the first scientific director of the new, independent SETI Institute. Then, and now, the institute operated fully on the basis of sponsoring and donations from the private sector.

DURING MY visit to the SETI Institute at Mountain View, Jill Tarter managed to make half an hour free in her diary to speak to me. Officially, she is retired, but you would never know it. It is October, deadline season. Observation proposals and funding applications have to be sent off. She is still in a meeting. Tarter's assistant, Chris, shows me into her office. The walls are full of photographs of donors, friends and admirers. Of course, the signed photographs are the first to attract my attention. The famous picture of 'Earthrise' from the Moon, signed by the crew of *Apollo 8*. A picture of Jodie Foster, whose role in the film *Contact* is based on Tarter, with a personal message of best wishes. Above her desk hangs a gigantic portrait of Jill herself with hundreds of signatures of friends and colleagues, made to mark her 'retirement'. Chris shows me another photo, for Tarter the prize of the collection. It is a shot of all SETI staff on a cruise ship in Alaska. The ship is owned by Paul Allen, co-founder of Microsoft and one of the institute's largest donors. The other donors include many well-known names, such as Intel founder Gordon Moore, computer manufacturers Hewlett and Packard, and science-fiction author Arthur C. Clarke.

Then Jill Tarter comes in. Short, grey hair, comfortable but stylish clothes, and a friendly but penetrating look in her eyes. She looks far too young to be retired. She apologizes for her busy agenda and asks me first in detail about my research project. Then we talk a little longer than a half hour about her life's work. About how she first became acquainted with the Cyclops report, about the primitive and very heavy PDP-8/s desktop computer ('Oh, it was just so funny,' she whispers to herself), about how she recruited Seth Shostak at a party in Berkeley, about the many projects that have seen the light of day partly due to her efforts.

One of those projects is Serendip, in which SETI experiments hitch a ride with other astronomical observations. That gives SETI more valuable observation time. Popular telescopes are so heavily booked up that it is like interviewing Scarlett Johansson, when journalists are permitted to ask only one question. Tarter is also involved in the crowdsourcing initiative SETI@home, in which everyone with a computer can take part from their own living room. She is also contributing to the Hundred Year Starship, an ambitious plan to build a spaceship in the coming hundred years to fly to the nearest stars.

Tarter also tells me about the criticism that has been aimed at SETI for more than fifty years. The 'little green man' stamp has never completely gone away. Sceptics also point out that the success of the experiments depends on the extent to which aliens resemble humans. Prominent physicist Freeman Dyson, for example, expresses the reservation that SETI experiments assume that there are civilizations which, like us, broadcast codable messages. 'Since it may be untrue,' says Dyson, 'I am in favour of looking for other evidence of intelligence, and especially for evidence which does not require the cooperation of the beings whose activities we are trying to observe.' A remark like this is reminiscent of Christiaan Huygens, who noted three hundred years ago in *Cosmotheoros* that alien life could very easily 'dwell in any other shape than ours'. Dyson does, however, admit that since he cannot prove either that communicating aliens do *not* exist, he is 'whole-heartedly in favour of searching for radio messages'.

The recent discovery of planets outside the solar system has given the SETI Institute a significant boost. Yet the hard truth is undeniable: they have been looking for fifty years and have found nothing at all. The most remarkable result in that whole period was the 'Wow!' signal. In 1977, a radio telescope at Ohio State University picked up a radio signal that was exceptionally strong and could not be explained. 'Wow!' wrote the duty observer on the computer printout. But the signal was never picked up again.

Jill Tarter is passionate in responding to criticism. She emphasizes that space is so large that we can hardly comprehend how minuscule our efforts have been so far. Anyone who sees that in proportion will realize that it is not at all peculiar that we have not yet found E.T. 'The ocean is teeming with fish, but if you scoop a single glass of water from it, the chances that there will be a fish in it are very small.' Not to mention a dolphin. Her advice is therefore to use a bigger glass.

Despite the incontestable logic behind Tarter's explanation, the success of the SETI experiments is not the most important aim. That was clear to me after spending a day at the Institute. The chances that we ever receive a radio signal from E.T. cannot be estimated with any degree of accuracy. Perhaps it only happens in films like *Independence Day*, and maybe we should just leave it at that. Those aliens weren't exactly friendly. It is probably quite unlikely that there really are space creatures that have built radio equipment. On the other hand, imagine that they do exist. The impact of that discovery would be inconceivable.

The most valuable thing that SETI gives us is inspiration. That is what drives Jill Tarter, Seth Shostak and the hundreds of scientists conducting experiments around the world. Crossing borders, doing what no one ever considered possible. Reaching for the stars. Tarter says about her participation in the Hundred Year Starship project, 'I've signed up because I'm a big fan of Arthur C. Clarke. He said "The only way of discovering the limits of the possible is to venture a little way past them into the impossible." I don't know if it is possible to make an astronaut travel between the stars at even a fraction of the speed of light. But I'll never know the answer until we've tried it. Do you know what the last sentence of Cocconi and Morrison's article was? "The probability of success is difficult to estimate; but if we never search the chance of success is zero." That has been SETI's motto for more than fifty years.'

We are already five minutes over my allotted time. The telephone rings. As I leave, I hear Tarter talking to a colleague. 'Yes, the observation proposal looks good,' she says. 'Maybe we can search on a broader frequency if we use the 1422 MHz filter . . .'.

A planet hunter never retires.

five

THE TENACIOUS INVENTOR

I WAS RECENTLY leafing through an old, dog-eared encyclopaedia that I used to read over and over again as a boy. The green cover was half hanging off, and the pages were musty and faded. I found my favourite page, about the Apollo missions to the Moon. It was a comic strip and the pictures were spectacular: the launch of *Apollo 11* in 1969, the separation of the different rocket stages, an image of the Earth from space, the landing on the Moon, Neil Armstrong's first footprint. I still know every detail of that page from memory.

The final picture, of the astronauts returning to Earth in their tiny capsule, fascinated me the most. As the capsule re-enters the atmosphere, the air resistance makes the outer shell so hot that it glows like a fireball. I have always wondered how on Earth they did that. How did they know beforehand that the astronauts in the capsule wouldn't be burned alive? How thick did the shell of the capsule have to be? And what was it made of? How did they make sure the astronauts returned to Earth safely?

'LUCAS? HELLO, I'm Bill Borucki.'

Twenty years after first reading the encyclopaedia, I meet the man who knows the answers to my questions. He is a research scientist who helped design the heat-resistant shields for the Apollo capsules. In front of me stands a small, somewhat older man with bushy eyebrows. His eyes look alert under his golf cap. We order an Indian Pale Ale and drink it on the terrace of the

Shoreline Golf Club in Mountain View, a few kilometres from the Google campus. We are the only ones on the terrace. 'I don't play golf,' Borucki tells me, 'but I like to come here because it's quiet.' We enjoy the afternoon sun and the breeze from the ocean. The only thing that occasionally interrupts our conversation is the loud honking of a flock of geese.

Borucki speaks softly and fast, swallowing his words a little, but in elegant full sentences. Sometimes, I miss the jokes he drops in here and there, as he delivers them completely straight-faced. He describes the need for the heat shields in simple terms: 'The astronaut is sitting in a spaceship heading for the Earth at 40,000 kilometres an hour. Half a metre from his butt, it is 1,000 degrees hotter than the surface of the Sun. If the astronaut heats up above boiling point, he dies. My job was to get that butt home safely, okay?'

Borucki was fascinated by making and firing off rockets from a very young age. He used to make them with his brother and their friends. At first, they used match-heads to launch them, and later they mixed their own gunpowder. Sometimes, one would explode. There is a picture of him as a boy of about sixteen, standing in a field in his home state of Wisconsin. He is wearing a lumberjack shirt, tucked into trousers that are pulled up high around his waist. Short hair in a neat parting, high cheekbones and dark eyebrows. In his hands, he is holding a small rocket. A conspiratorial smile plays on his lips – he's clearly about to launch the thing skywards.

One of Borucki's boyhood hobbies was gazing at the stars. He lived an hour's bicycle ride from the renowned Yerkes Observatory, where Otto Struve started his career and of which he was director for many years. In the summer, Borucki would sometimes go there with a friend to look through the telescope, at that time the largest refracting telescope in the u.s. On other summer evenings at home, he would lie on his back on the roof and watch the shooting stars of a meteorite shower. In the daytime, he would build radio equipment to communicate with extraterrestrial life, a kind of SETI *avant la lettre*.

Although he studied physics in Madison, the capital city of Wisconsin, he never lost his passion for astronomy and making things. At that time, NASA was recruiting people of Borucki's calibre en masse for the Apollo project. The space agency badly needed young talent, as it wanted to send men to the Moon but had no idea how the laws of nature would behave in the extreme circumstances of space.

Borucki even received two offers. 'You didn't have to be a straight-A student to join the team,' he tells me. 'You didn't even have to write a dissertation. They had no time to work out elaborate theories. It was a race to beat the Russians to the Moon. That was the thought that drove them, kept them going. And they were looking for people who could find quick and practical solutions to problems. I was someone like that. I love putting my ideas into practice. The team had as much money as was needed to overcome the obstacles in their way. We just had to get it done.'

So Borucki found himself in his element, at an enormous ballistic test range at NASA's Ames Research Centre. This was the test lab for the heat-resistant shields that had to protect astronauts' butts when they returned to Earth. In true Borucki style, the approach was no-nonsense and straightforward. Two cannons were set up, facing each other. One, a series of naval cannons linked together, fired off 75 kilograms of compressed air at 5 kilometres a second. The other shot a scale model of a space capsule through the air in the opposite direction. The friction caused an ultra-hot shockwave on the leading edge of the model spacecraft. 'Those guns are still at Ames,' Borucki confides to me, a mischievous look in his eyes. He waves his arms around in the air to describe the spectacle: 'Ha! That tiny spaceship . . . It shot out of the barrel of the cannon at such a speed, pushing a plasma hotter and brighter than the Sun and enveloped in material burnt from its surface. You saw the shockwave bouncing off it! Sometimes, the whole thing had evaporated before it got to the end of the test course. It was temperatures like that we had to protect the astronauts from.'

Borucki's job was to design equipment to analyse the light from the shockwave and calculate the temperature of the air around the capsule, to test the performance of the heat shield. Since so little was known about the atomic properties of air at such high temperatures, the measurements were compared to those of bolts of lightning, which reached even higher temperatures. The extreme tests and observations enabled the team to calculate the temperature of the air around the spacecraft, so that they could design heat shields from insulation material that could withstand such heat. That was the hard reality: build a model, test it and adapt the design until it worked. And fast, before the Russians did it.

The shields worked. 'Of course they worked,' says Borucki. 'NASA had doubled our safety margins – and we had doubled them ourselves.' The Americans were the first to land on the Moon and the astronauts came back with their butts intact, but three years and six successful missions later, in 1972, the programme was wound up. 'They then fired our team and told us to find other new and important problems to be solved,' Borucki tells me.

Borucki joined the Theoretical Studies department at Ames, where he studied the atmosphere of the Earth and the other planets in the solar system. The research could be used to address a lot of practical problems, like the impact of greenhouse gases on the climate. 'It was always a disappointment that people – and especially government leaders – do not wish to understand the impact of the greenhouse effect,' said Borucki. 'The principle is very simple. CO_2 in the atmosphere warms up the Earth. It acts like a blanket. If your body already generates heat and you wrap yourself in a blanket, you will warm up. You don't have to be an intellectual powerhouse to understand that.'

Another phenomenon that Borucki studied was thunderstorms on other planets. The space probe *Voyager 1* and other detectors had picked up light flashes and radio emissions on Jupiter, Saturn and Venus which were identified as bolts of lightning. But Borucki was interested in lightning for other reasons, too. Not only for heat-shield design, but because lightning produces chemicals in the atmosphere that destroy ozone and can affect the Earth's climate. Furthermore, lightning in planetary atmospheres can produce molecules that might lead to the evolution of life. This helped find answers to important questions like how life had originated on Earth and whether it was also possible elsewhere in the solar system.

THERE HAD long been speculation about the role of electricity in the origins of life. Electricity has something magical: it can make lifeless things move. Lightning, an electrical discharge from the sky, still sparks our imaginations. In the book *Frankenstein*, written by Mary Shelley in 1818, a young scientist sees lightning strike an oak tree. He wonders whether this destructive power can be used to create life. He uses electricity to bring life to a monster that he has pieced together from dead body parts, with catastrophic consequences.

Nineteenth-century writers of non-fiction also wondered whether electricity had played a role in the emergence of life. Charles Darwin stated in his theory of evolution that all life forms on Earth are constantly developing. In 1871, Darwin told his best friend, the botanist Joseph Hooker, that all the different species of animals and plants on Earth could eventually be traced back to a single primitive ancestor. He told Hooker how this form of life had emerged: 'But if (and oh what a big if) we could conceive in some warm little pond with all sorts of ammonia and phosphoric salts, light, heat, electricity etcetera present, that a protein compound was chemically formed, ready to undergo still more complex changes . . .'. Could life, Darwin asked himself, have been created in that small, warm pond?

Darwin's recipe for life was, in short, heat up a couple of litres of primal soup, zap it with a few bolts of lightning and wait for a while. Darwin admitted that he did not have the means to prove his hypothesis. He also had no idea at all what kinds of chemical reactions would lead to the emergence of living organisms. But his main idea of a small, warm pool is not that far-fetched. Chemical reactions set in motion by electricity can produce simple building blocks from which complex structures are formed that in turn can lead to the development of animals and plants.

This was proved by the 'primal soup' experiments conducted by Stanley Miller and Harold Urey. In the 1950s, they put Darwin's ideas into practice at the universities of Chicago and San Diego. The experiments would not have been out of place on *Masterchef*. The chemists made a cocktail of a lot of different chemicals that were present in the early atmosphere, including water, methane and ammonia. Then they heated it up, fed it through a system of tubes and zapped it with electrical sparks to simulate lightning. They repeated the process several times. After a day, the liquid had turned pink and, eventually, there was a brick-red sediment left over, in which a number of permanent solids had formed. Among these, the researchers found about a dozen different amino acids, the complex molecules that make up protein. Amino acids are also the elementary elements of DNA. It was a long way from being living material – the difference between this sediment and a living cell was comparable to that between a copper wire and the Internet – but Miller and Urey had shown that an electrical current could make simple molecules react to form complex molecules.

BORUCKI SUSPECTED that Miller and Urey's results also applied to other planets. Could amino acids be formed there during thunderstorms, too, in an atmosphere that was different to that on Earth? He wanted to work out how much energy was generated by lightning on Jupiter and the other planets. And then he wanted to know how much of that could eventually be used to create molecules. The only way to find that out was, of course, by experiment.

The aim of the experiment was to make lightning in the laboratory, in a gas that was similar to the atmosphere of a planet. Using a spectrometer, Borucki discovered what changes the lightning caused in the atmosphere. As with Miller and Urey's experiment, he was able to measure what molecules were formed. Lightning simulation, using electrical sparks, was already widely used in aviation tests. 'If lightning strikes a plane hard,' he says, taking a sip of his beer, 'it can cause a fuel leak, or a wing can break off. It's not good if a wing breaks off.'

There was a practical problem with Borucki's lightning experiment. 'We weren't talking about lightning on Earth,' he said, 'but on Jupiter. Jupiter's atmosphere is made up of hydrogen gas. If you duplicate that in a laboratory and the hydrogen leaks, the laboratory will blow up. That's not a good thing. People don't like it if you blow up their lab.' Lasers offered a possible solution. Instead of an electrical spark several metres long, Borucki's team wanted to use a focused laser to generate a bolt of lightning in a small bottle. That would only mean using a small quantity of hydrogen gas and any explosions would be limited in scale.

Borucki sighs. 'Of course, there was no money available for it, because it had not yet been proven to work,' he says. So he had to conduct the experiment outside office hours and with the materials he had at hand. He and a lab assistant who also liked to experiment met at the laboratory on Saturday mornings. It was just as well that the lab was empty, as the laser made a terrible racket. But it worked. They succeeded in generating small lightning-like discharges safely in a test tube. Borucki wrote a research proposal and received funding to continue the experiment. When a NASA headquarters scientist saw the proposal, he said that the experiment wouldn't work and shouldn't have been funded. 'But by that time our team had simulated lightning in the atmospheres of Jupiter, Venus and Titan, and published the results in a professional journal. Headquarters replied "Oh, okay, never mind."'

This is how breakthroughs happen – by people tenaciously putting their ideas into practice, no matter what their colleagues say. 'A good experiment is one that most people say can't be done,' says Borucki. 'They say "Damn it, Borucki, it's not gonna work! Get outta here, go do something useful!" The number of times I've heard that. You have to do your experiments while no one is looking, because everyone knows it won't work. And who knows? Maybe they're right.' Even before his work on planetary atmospheres, Bill Borucki had attended the SETI workshops at Ames in the early 1970s. These were the workshops where they explored how best to estimate the factors in the Drake Equation. Where they first used the term 'planet hunters'. And where a biologist asked the simple question, how often will Earth-like planets form around stars other than the Sun?

The serious astronomers around the table couldn't give the biologist an answer. There were no hard and fast observations of planets. The only serious candidate, Peter van de Kamp's planet, had been successfully torpedoed by his successor, Wulff Heintz. Otto Struve's suggestion, based on observation, that planetary systems must occur frequently, was only one of the many ideas doing the rounds. Another popular theory, for example, was that the Sun had once collided with another star, and that the Earth and the other planets had been formed from the debris. That would make the solar system unique.

With hindsight, it would seem logical for planets around other stars to be the rule rather than the exception. After all, why should we be special? The Sun appears to be a completely regular star, so why should it be the only one with planets? Forty years ago, however, this was anything but the accepted truth. The lack of data left the way free for exotic theories. The curious were short-sighted: observations and measurements were not yet technically possible and few believed that would change in the foreseeable future.

In the 1960s and '70s, attention was focused fully on disciplines where observations and theories did run in parallel. Cosmology, the study of the origins of the whole universe, became increasingly popular as more and more of the universe was explored and charted. Experimental physics was also on the rise, as new elementary particles were discovered by the dozen. This allowed different theories on the nature of matter to be confirmed or rejected.

As Jill Tarter had said before, planets were simply not sexy. The lack of data meant that theories went completely off the rails. Far-reaching and overly detailed speculations about aliens showed just how much science had been hijacked by science fiction. Efforts to observe exoplanets were still performed only by a handful of old-school astronomers. In the absence of theoretical work to counterbalance their efforts, they were ignored or – in the case of Van de Kamp – negated, which did not promote the credibility of planetary science.

Serious scientists with an interest in the extraterrestrial, like the members of the Order of the Dolphin, breathed new life into the discipline. With their 'crowdsourcing' aspect (everyone could take part from their own back gardens) and their adventurous nature (searching for extraterrestrial life), SETI and planet-hunting were very popular among amateurs, philanthropists and academics crossing over from other disciplines. In 1983, while attending the workshops at Ames, Bill Borucki stumbled across an article that had been written twelve years previously by an author from the latter category, the academic adventurer Frank Rosenblatt.

ROSENBLATT WAS a professor at Cornell, where he had studied psychology. After getting his PhD, he became a pioneer in the development of artificial intelligence. He invented the Mark I Perceptron, a primitive computer that could teach itself simple tasks, such as distinguishing between a triangle and a square. He also developed a theory about how the brain works and transplanted a rat's brain to show that memory is not physically transferable. And, as if that were not enough, Rosenblatt was also an accomplished pianist. His endless improvisations of *Three Blind Mice* were especially renowned on campus.

On top of all this, Rosenblatt was a keen amateur astronomer. With the help of his friends and students, he built an observatory on a hill behind his house. Astrobiologist Carl Sagan was a colleague at Cornell and Rosenblatt became interested in SETI. He thought about how to solve Drake's Equation, especially the question about how many planets stars have on average. He came across the 1952 article mentioned earlier, in which renowned astronomer and SETI proponent Otto Struve proposed two methods for detecting planets. Rosenblatt worked one of these out in detail: the transit method, by which planets can be detected as they move in front of their stars.

Just imagine you're flying in a spaceship so far away from the solar system that the Sun is only a small dot. As long as you keep flying in the same plane as the Sun and the Earth's orbit, you will see the Earth pass across the face of the Sun once every year. You will also see the other planets do the same thing, each according to their own orbital period. During its transit, a planet blocks a minuscule part of the light emitted by the Sun. The planet itself is too small to distinguish separately, so its presence has to be measured by the strength of the Sun's light. If the Sun appears to burn slightly less brightly and this occurs at regular intervals, we can conclude that a planet is orbiting it, eclipsing its light a little with each transit. From the Earth, we can see the transits of the two planets closer to the Sun than we are, Mercury and Venus. We saw earlier how Captain James Cook observed the transit of Venus in 1769, making it possible to calculate the distance from the Earth to the Sun.

Struve's idea was to observe these variations in the light emitted by other stars. If a star regularly became a little fainter, it could mean that a planet was moving in front of it. However, a planetary transit is only visible from Earth if the planet and its star are in the same plane. As this is not the case with the majority of stars with planets, we rarely see exoplanets move in front of their stars.

Rosenblatt proposed keeping track of a large number of stars at the same time. There was sure to be a few among them positioned so that we could observe a planetary transit. Their brightness would have to be measured very accurately, as a star's light is reduced by only a very tiny fraction during a transit. This fraction, also known as the 'depth' of the transit, depends on the size of the planet relative to the star: the larger the planet, the more light it captures. Once the size of the star is known, which can be worked out quite easily, the absolute size of the planet can also be determined. By applying the laws that Kepler devised to describe our own solar system, the fixed interval between each transit indicates the distance from the planet to the star. That, in turn, gives the temperature on the planet's surface.

Rosenblatt worked out how many stars would have to be monitored at the same time to find at least one planet. Like Struve, he took our own solar system as an example and assumed that most stars would have the same kind of planetary system. Based on this and other assumptions, he deduced that, to detect one Jupiter-like planet a year, some 9,000 stars would need to be monitored continually. One of the challenges of the project was the enormous computer power required to process and save all the measurement data. 'A storage capacity of 50,000 to 100,000 words seems to be needed,' Rosenblatt wrote, with some concern. That is approximately the number of words in this book. Ten such books would fit on a floppy disk, but floppies had not yet been invented.

WHEN HE read Rosenblatt's article in 1983, Bill Borucki was impressed. 'A large-scale monitoring campaign was a smart and new idea,' he says. 'There were a few errors in the article, but that was no problem. You can solve errors; that's what they're for.' When Borucki tried to contact Rosenblatt, he heard that he had died several years previously, suffering a fatal accident while kayaking in 1971, on his 43rd birthday. The article had not been published until after his death. So Borucki was on his own but, thanks to Rosenblatt, he now knew how to find exoplanets. 'All I had to do was develop instruments to enable me to do that,' he tells me.

That instrument did not yet exist. Photometers, which measure the brightness of stars, were nowhere near powerful enough to register a planetary transit. If a planet as large as Jupiter passes in front of a star, it will reduce the light emitted by the star by around 1 per cent. This minuscule dip in luminosity would be even smaller if you are looking for an Earth-like planet, because that would be a hundred times smaller. But that was exactly what Borucki was looking for, a twin planet of the Earth on which life would in theory be possible. 'If we wanted to find planets, which ones would be interesting?' Borucki says. 'Screw Jupiter-sized planets. Planets like that, which consist only of gas, cannot support life. The only ones that count are rocky, habitable, Earth-like planets. Planets that are the right distance from their star, so it is not too hot and not too cold. That is the step we wanted to take. To answer the question, are habitable planets common? If the answer is yes, the universe is probably teeming with life. If it is no, none of it matters. You can build a great spaceship, like in *Star Trek*, but there's no place to go.'

Borucki worked the plan out in detail, together with computer programmer Audrey Summers. In 1984, they published an article in which they examined Rosenblatt's results. In a somewhat judgemental tone, they stated that 'Rosenblatt's estimates are too optimistic' and that his article needed some correction ('equation 1 has to be swapped with equation 12'). After doing Rosenblatt's calculations again, Borucki and Summers came to three important conclusions.

First, they calculated the number of stars that would need to be monitored simultaneously. Like Rosenblatt, they proposed designing a telescope that would continually monitor the same section of the sky. If all stars had a planetary system like ours, they estimated that a handful of Earth-like planets could be discovered. If, on the other hand, they didn't find anything, that would mean that other stars generally speaking have no planets. Our solar system would be an exception to the rule and we might be alone in the universe. To reach that conclusion with any certainty, Borucki and Summers felt that it was necessary to monitor at least 10,000 stars, continuously and simultaneously.

Their second conclusion was that, to register the transit of an Earth-like planet, they would need a stable photometer that could simultaneously register the brightness of thousands of stars with an accuracy of ten parts per million, a degree of precision that had never been achieved before. Imagine that the light from the star is equal to the combined light from all the floodlights in a football stadium. The tiny fraction of light absorbed by a planet would be roughly equivalent to that of the bulb in a bicycle light. The photometer that Borucki and Summers were proposing therefore had to be able to measure from an enormous distance whether a football supporter had his bicycle lamp switched on during a floodlit match.

Borucki grins as he recalls the responses of the photometrists that he asked to help him build an instrument that powerful. 'Oh, it was hilarious,' he says. 'At a certain point, most of them would not talk to me anymore. They thought I was really losing my marbles.'

Lastly, Borucki and Summers said that the desired result could never be achieved with a terrestrial telescope. Stars always seem to flicker a little, even on clear nights. These fluctuations are caused by the turbulent mixing of parcels of warm and cool air in the atmosphere. The transit of an Earth-like planet orbiting another star is much too subtle to observe and will always be overwhelmed by the interference in the Earth's atmosphere. But that problem does not exist in space.

So Borucki's wish list was pretty ambitious: he wanted to mount an instrument that did not yet exist on a satellite he hadn't yet got and send it up into space on a rocket to continually monitor 10,000 stars with a precision that had never yet been attained on Earth, to find planets for which there was no evidence at all of their existence. It was a mission impossible. And yet Borucki was determined. 'You only get one chance,' he says. 'Forty years or so and' – he slaps a hand down on the table – 'it's over. Then you will ask yourself, what have I actually done with my life?'

THE ARTICLE that Borucki had written with Summers was completely ignored. A year later, he published another one, with two other colleagues; that was ignored, too. 'At the presentations I gave to photometrists and other astronomers,' he tells me, 'everyone agreed that my proposal would never work. It was rejected.' It probably didn't help that Borucki was seen as a bit of an odd man out. He wasn't an astronomer. He had no experience with the kind of photometers he wanted to build. He had never got his PhD, and some scientists looked down on him for that. Once, a teaching invitation was cancelled because he didn't have a 'doctor' before his name.

With great difficulty, he managed to extract some money from Ames to organize workshops, where he brainstormed with experts on possible designs for a ten parts per million photometer. It was customary at that time to use photomultiplier tubes to make precision measurements. Borucki knew that they would never achieve the required degree of precision. 'But the experts were convinced that photomultipliers were God's own creation,' he says. He contacted a few radical figures who were experimenting with CCD detectors.[18] Today, CCDs are used in all optical equipment, from telescopes to digital cameras. Back then, the technology was new but, in the 1990s, Borucki's team succeeded in making a prototype that could achieve a precision of 10 parts per million.

Yet the astronomers to whom he showed his CCD results remained sceptical. His new boss, Dave Morrison, took him to one side and said: 'Bill, everyone agrees that a space-borne photometer will not work. But . . . I'll give you one last chance. I'll put together an independent committee of scientists and you can present your ideas to them. If they say yes, you can carry on. If they say no, it's curtains for the project for good.'

The committee was chaired by Jill Tarter who, like Carl Sagan, supported the search for exoplanets. Other than Tarter, support for Borucki was thin on the ground. A few heavyweights from the astronomical world were called in to torpedo the proposals. 'Poor Bill. He had a lot to defend,' Tarter recalled. The main concern of these big shots was that the brightness of many stars already varies as a result of internal processes. It would be impossible to distinguish the transit of a small planet from these normal fluctuations. They were also sceptical about the modern technology of the CCDs. In a sweaty session lasting several hours, Borucki took all their blows on the chin and patiently explained why it *would* work.

And with success. By the end of the interrogation, he had persuaded a majority of the members that the plan was feasible. Dave Morrison approved the further work and found funding for it. Some of the committee members, including Jill Tarter, joined his team as advisers. Borucki was able to continue his work on developing a space mission to find exoplanets. David Koch became his right-hand man. After Koch died in 2012, Borucki described him in an obituary as 'the only other person who believed in this mission for many years'. Besides being an exceptionally gifted instrument-builder, Koch was also someone who kept others' spirits up. He made caps for the other members of the team, was well-known for his corny jokes and cheered his colleagues up with cartoons when, yet again, they were confronted with setbacks.

KOCH'S JOKES were welcome, as there were setbacks aplenty. In the early 1990s, NASA announced an open competition for a satellite mission. The winning plan would be given money to build a telescope, which would be sent into space on a satellite. The competition would be held every two years, each time with a few dozen candidates.

Borucki's team entered the competition in 1992. They tested a new prototype CCD detector by shining lamps through a metal plate with small holes in it, representing the stars. The tests showed that the CCD had the precision required for the mission to succeed, but the proposal didn't make it past the first round. '"Such detectors don't exist," the panel told us when they saw our plan,' Borucki sighs. 'But they most certainly did. We'd already made them, but we hadn't had enough time to write an article about them.'

When they submitted the plan again in 1994, it received support from an unexpected quarter. In 1992, astronomers had found a number of planets around a pulsar (a neutron star like those Jocelyn Bell had discovered in the 1960s) using a method based on irregularities in the pulse period. They were not the kind of planets that Borucki was looking for – orbiting stars like the Sun – but confirmation that such objects existed was a welcome boost. But, this time too, the proposal was rejected. The panel felt that there was still insufficient evidence that the detector worked. Furthermore, the mission would be too expensive. 'It was nonsense, but we simply had no time to process and publish our test results,' Borucki says. 'The work we did on the plan was not a full-time occupation. NASA expected you to do something useful with your life – all the while I was mainly busy working on other space missions.'

Borucki was seriously disgruntled about the rejection for a few days, but then he stoically picked up the thread again to make a fresh start and prepare for the next competition two years later. This time, an important part of the plan was changed – its title. Until then, the proposed mission had been known by the acronym FRESIP, which stood for Frequency of Earth-sized Inner Planets. 'It was a completely logical name,' Borucki said, defensively. After all, he had thought of it. 'We wanted to know how frequently Earth-sized planets occurred. In our own solar system, they are the closest to the Sun, so that's why we call them inner planets.' But other team members, including Koch, Tarter and Carl Sagan, who had by this time also joined the team, thought FRESIP wasn't catchy enough. 'They said "Bill, we've been rejected so often now. Shouldn't we change the name of the satellite? Let's call it Kepler. A famous astronomer, the first to describe the orbits of planets, and he did important work in optics." So I said, "Well, if you let go my arm before you break it, alright then." And so we changed the name to Kepler.'

Besides the change of name, there was another significant factor that improved the project's chances of winning. From 1995, as well as around pulsars, exoplanets had been found orbiting Sun-like stars. That made the Kepler mission credible. Exoplanets were no longer seen as only a fantasy in Borucki's head. The real hype was yet to come, but at least planet hunters were no longer completely ridiculed in scientific circles.

THE NAME Kepler and the discovery of the exoplanets undoubtedly had a positive effect on the panel in 1996. Borucki convinced them that the detectors worked and that the project would stay within budget. But, this time too, there was an obstacle: the number of stars that Kepler would have to monitor simultaneously. Until then, no single astronomer had observed more than one or two stars at a time. 'We claimed that our detectors could watch 10,000 stars at a time,' says Borucki. 'They said "Bullshit. Prove it. Build a real observatory, observe 10,000 stars and prove it." So we did just that.'

He contacted the University of California's Lick Observatory in the mountains of the Diablo Range, 50 kilometres from Ames. 'They thought that I was crazy to undertake such a project there, too, but they liked crazy types. They had an old, unused dome there that they said I could use.'

Borucki drove to Lick to take a look. The dome had not been used for seventy years and was a ruin. The round roof leaked. Rodents pattered across the rotting floor, and snakes slept underneath it. The dome could no longer rotate and the base for the telescope had ceased to exist. There was no toilet and nocturnal observers had to use one in another building, with mountain lions on the prowl outside. Borucki was delighted: 'This was the greatest gift I could wish for,' he says. 'If you want to build something new in the u.s., it takes years just to get the environmental report written to make sure that the red and green squirrels, the yellow tree frog and other weird concoctions of nature are not endangered. This observatory was already there. OK, the floor was rotten. But I had once built a house with my father. Patching the dome and replacing the floor shouldn't be a problem.'

The team set to work. They laid a new floor, and built a new telescope from spare parts. None of the components could cost more than $3,000 – the withdrawal limit on the credit card they had been given by NASA. They made a calibration lamp from a plastic bucket, and set up a large umbrella to protect the telescope from the rain. The result of all their efforts was christened the Vulcan Telescope, after a legendary, never-discovered planet in the solar system. The name had a double significance, as Vulcans are also an alien race from *Star Trek*. As a further tribute, the house mouse at the observatory was called Spock. With Spock as their mascot, the team was ready to make the fourth attempt to get their satellite off the ground.

But it was all to no avail. In 1998, the proposal was rejected for a fourth time. The observations with the Vulcan Telescope may have shown that the method and the equipment worked and that 10,000 stars could be monitored at the same time, but the jury still had one objection. How did they know that it would all work in space? Would the precision instrument be able to withstand the vibrations of the spacecraft?

So it was back to the drawing board. NASA gave the Kepler team enough money to build a new prototype within two years. They patched together a model at breakneck speed. It consisted of a long, closed box a little like a mobile toilet cabin, with a light at the bottom. Above the light was a metal plate with minuscule holes that produced a kind of artificial night sky when the light was switched on. Through an ingenious system of ultra-thin wires, some of the apertures could be made slightly smaller, to simulate a planetary transit. A telescope and CCD detector were positioned in the top of the box to capture the 'starlight'. The telescope was shaken back and forth to simulate the vibrations of space flight. Borucki's team then developed computer programs to compensate for the vibrations.

In 2000 they were ready and the proposal was submitted for the fifth time. 'We showed them that the detectors worked,' Borucki says. 'That we could now even monitor 150,000 stars from space at the same time and that the costs were manageable. And we won. That is to say, we were through to the next round.

We had a chance to compete with two other missions, rather than several dozen. We won that second round, too. They chose Kepler above a mission to Jupiter despite strong support for the latter. The people on the panel were gutsy enough to take a chance on our mission. I really admire them.'

After nine years and four rejections, their efforts were finally rewarded: they had been given the green light to start developing the Kepler mission. By now, the team had several dozen members and hunting planets was high on NASA's agenda. But their troubles were not yet over. NASA said there was too little money to start until the following year. In that first year, by pulling out all the stops, NASA managed to wheedle enough from several different funds to order a few important components. Then in 2005, they were told that the budget for that year would be halved. 'Because we had already spent half of the budgeted funds, we had to lay off many people,' says Borucki. A year later, when the budget was approved again, the contractor that was building the spacecraft took on and trained new people. It wasn't a particularly efficient way of working. In the years that followed, a streamlined management team took over the running of the project ('That was good,' says Borucki. 'I could concentrate on the science again'), and the scientific team expanded steadily.

On 6 March 2009, a week before the 25th anniversary of his first article on planetary transits, the day arrived that Bill Borucki had been waiting for all that time. The Kepler satellite was launched by rocket from Cape Canaveral in Florida. It was a clear, windless evening. 'They'd asked me if I wanted to sit in the control room to toast the launch with the scientists and engineers, to say, "Yeah, we did it,"' Borucki says. 'But I chose to be outside, with a good view of the rocket, and to be with my wife and kids and grandchildren. Life is always a balance.'

WE'VE FINISHED our Indian Pale Ales. Borucki and I are both quiet and we listen to the geese on the terrace. The man I have just spent two hours talking to is exactly as others had described him. Natalie Batalha, for example, one of the project leaders who has been working with Borucki for thirteen years, wrote in her blog: 'What left the biggest impression on me was Bill's constantly positive attitude in the face of rejection and criticism – his staunch resolve to never take criticism personally but rather to use it as a means of improvement – and his unwavering persistence.'

On the terrace, I ask him a question that has been running through my mind for a while. 'All those setbacks . . . It seems that they only encouraged you, rather than discouraged you.'

'Pissed off is a better way of putting it! How could they not understand it! I was always pretty sick for a week afterwards. Then the tirade was over and I said, OK, let's get the job done. I'm amused that people always write that I'm so tenacious. "Stubborn" was the word the management always used. But that's not how it felt to me.'

'What would you call it then?'

Borucki thinks for a moment and then says slowly, 'I'm dedicated to getting the job done. Getting to the answer, that's what counts. Get the answer, no matter the obstacles.' He bangs a hand down on the table. 'Damn the torpedoes, full speed ahead!'

FOUR MONTHS before my meeting with Bill Borucki in Mountain View, Kepler experienced problems. Two of the reaction wheels that keep the telescope constantly pointed at the same area of the sky had stopped working. It was impossible to repair them, but the satellite got a second chance, by using one of its solar panels to keep it in position. In true Borucki style, they solved the problem by improvising.

The mission is, however, already a success. As this book went to press in 2017, in its small piece of sky, Kepler had discovered nearly 5,000 potential exoplanets, over 2,300 of which have been confirmed as real exoplanets. This enormous haul has enabled astronomers to solve parts of Drake's Equation. Through statistical calculation of the planets discovered by Kepler, we now know that most stars have planets and that smaller planets are most common. The observation data sent back to Earth have brought about a revolution, an irreversible change in our knowledge of the universe. Exoplanets and those who discover them are in the spotlights.

When the Kepler satellite was launched into space in 2009, expectations were high. Borucki, who had started as a loner, now had the support of a small army of scientists who believed in his mission. Astronomers around the world waited with bated breath for the first results from the telescope. The seed of the revolution that Kepler had unleashed had actually been sown fourteen years earlier. A discovery had been made that more or less assured the success of the mission: the discovery of the first exoplanet around a Sun-like star.

Transit method

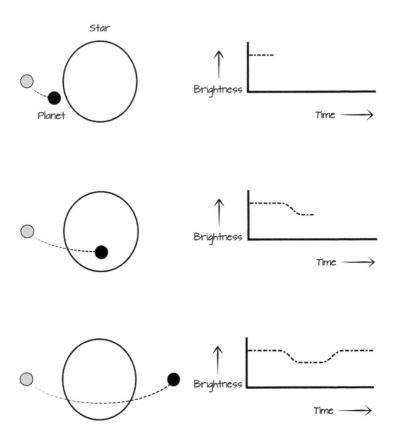

If an exoplanet passes in front of its star, an observer will temporarily receive less light from the star. This will repeat itself every time the planet has completed an orbit. The difference in brightness (the 'depth' of the transit), the duration of the transit and the orbital period make it possible to calculate useful properties, such as the relative size of the planet and the star, and the distance between them.

Doppler method

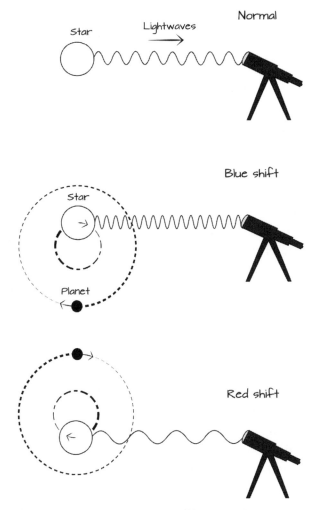

Planets are held in orbit by a star's gravity. But the planet's gravity also has a slight impact on the star, so that they actually orbit each other. The motion, or 'wobble', of a star can be observed by means of the Doppler effect. Light from a star that is getting closer looks a little bluer because the lightwaves are compressed, while light from a star that is moving away seems redder because the lightwaves are becoming more extended. This is known as blue and red shift, respectively. The duration and size of this colour shift indicates the relative mass of a planet and a star, and the distance between them.

six

A PLANET IN PEGASUS

O N THE morning of 10 October 1995, the professor was driving along the long, winding road. He had every reason to feel emotional. He had spent four long nights at the observatory on Mount Wilson, a mountain-top near Los Angeles. And for four nights, he had observed the same star. There was no other conclusion possible: the report from Italy was right.

He had previously scrapped the star from his list. It was different in chemical composition to the Sun; it wasn't the kind of star around which he expected to find planets. The professor was only interested in stars that did resemble the Sun, and he had more than a hundred on his regular schedule. Every night he drove to the telescope to observe them. He had a research programme that had already been running for twelve years, the last seven of which he had spent on non-stop observation. His aim was to see at least one of the stars make a movement that would indicate the presence of a massive planet.

He didn't expect to be able to confirm such a movement for perhaps twenty or thirty years: after all, the most massive planet in our solar system, Jupiter, had an orbital period of twelve years. Before the world would believe that he had discovered a planet around another star, he would have to observe at least two orbits. Until now, he had not seen any of the stars on his list make any suspicious movements. But he held on tightly to his long-term perspective: the secret was to keep observing patiently. His colleagues were sceptical, to say the least. But the professor kept his sights on the horizon and steadfastly went on observing, observing and observing even more.

Five days earlier, a colleague had called him from Florence. What he told him had given him four sleepless nights. 'Have you heard the news?' he had said. 'Two Swiss astronomers at a conference here claim they have discovered a planet around the star 51 Pegasi. Half the size of Jupiter. With an orbital period of four days. Four *days*, not years. Do you know anything about that?'

It couldn't possibly be true. It completely defied all the predictions. And there had been earlier 'discoveries' of planets that proved false. For hundreds of years, it had been assumed that our solar system was the standard for all other stars. If they existed, large planets around other stars would take tens of years to complete an orbit. Like Jupiter, which has an orbital period of almost twelve years. The Swiss astronomers must have made a mistake.

The professor did not hesitate, and drove straight to the mountain. He had four nights of observation left over from his last programme. If the Swiss claims were true, he could observe exactly one orbit of the planet around 51 Pegasi. His expectations were low. Surely everything that scientists had agreed on for so many years couldn't be wrong? Surely his strategy, which was to give him the world premiere as the first to discover an exoplanet, couldn't have failed? Could it be possible that a couple of Swiss researchers – the fact that they were French-speaking perhaps made it even worse – had beat him to it?

And the star had moved. In four nights, it had moved back and forth. That could only mean that there was something rotating around it, something half the size of Jupiter. The professor realized immediately that the Swiss astronomers were right. They had discovered the first planet outside our solar system. It was watertight.

He had been scooped. They had got there first and whipped the prize from right under his nose. He had devoted his career to this discovery and now he had come in second. He had not discovered the first exoplanet. You would at least expect him to spit out a curse or give his a car a good kick. But, eighteen years later, Geoffrey Marcy remembers exactly how he felt as he drove down from the mountain that morning: 'I felt delighted.'

AT 7.30 am I'm already in the coffee bar where we've arranged to meet. Geoffrey Marcy is the first astronomer to agree to meet me before 10 am. He is one of the most productive planet hunters in history, with more than 250 new planets to his name. A few days before our meeting, Reuters reported that he was on the shortlist for the Nobel Prize for physics. I have already finished my coffee when Marcy arrives. He is short, well-tanned and has a grey goatee beard. He's wearing a tennis shirt under a brown suede jacket, trainers and carries a sporty rucksack. He has a boyish face, but the bags under his eyes betray his busy job. Marcy is married without children. 'Planets are my children,' he says.

He orders a large cup of coffee and talks animatedly to a female student in a Berkeley tracksuit and carrying a tennis racket. Besides hunting planets, Marcy is also an avid tennis player. He will go to the court immediately after our interview, as he does every day. I ask him where he finds the time. 'I'm always working,' he says. 'Always. Four hours ago, I was still here, finishing off an article.' I am not surprised. One of Marcy's colleagues told me he had had a work appointment with him at 8 pm one Saturday evening, and that the next caller was at the front door at nine.

We walk across the beautiful Berkeley campus to his office. Neoclassical buildings with columns and tympanums stand amidst the fir trees, with paths winding between them. Students on their way to the first lecture of the day proudly wear their Berkeley clothes, with their imprint of Californian bears. MacBooks hum under the weeping willow. We come to the astronomers' temporary accommodation, a kind of Portakabin – even in idyllic Berkeley, as elsewhere, the natural scientists get the ugliest buildings. I notice that Marcy likes to chat to the students. This time it's a lanky young man, who is walking across the campus with a green cycling helmet.

'Wow, that's a great helmet!' Marcy exclaims. 'Where'd you buy it?'

'Er . . . at the bike shop?' the student replies, somewhat puzzled. He's probably never met this exuberant man before in his life.

Marcy's office in the Portakabin is small but pleasant. The bookcases are full of popular scientific books on exoplanets, with small models of rockets and planets in front of them. I sit on a couch that is too low and Marcy on his desk chair. He starts to talk about his childhood in Los Angeles. He speaks very animatedly, using big words and waving his arms around. Now and again, he takes a sip from his litre beaker of coffee. 'My father built airplanes,' he told me. 'He built parts of the Space Shuttle and used to tell me about that. Since then, I've had my eyes fixed on the stars.'

His parents had bought a telescope, which soon found its way to Marcy's room. Night after night, like Christiaan Huygens three hundred years earlier, he observed the motion of Saturn's moon Titan. He was given a subscription to the monthly magazine of the Griffith Observatory, an amateur observatory in the hills above Hollywood. The magazine was full of black-and-white pictures of the universe: of the planets in our solar system, of the nebulous regions where stars are born, of the spiral-shaped galaxies comprising billions of stars. 'I was amazed that the universe had so many jewels,' he said, spreading his arms. 'So much beauty. And that it was so big, so unimaginably big. The universe is *everything*. Psychology, anthropology, music, art – the study of the universe embraces all of that! As a child, I thought, why should you only look at a small part, if you could study the whole thing?'

Marcy went to the University of California in Los Angeles to study physics. 'I loved astronomy, but I had no idea you could make a living from it,' he told me. In the early 1970s, experimental physics was very popular. Study of the smallest (elementary particles) and the largest (the origins of the universe) dominated the agenda. It was the time when planets were not sexy and when Bill Borucki and Jill Tarter were just setting out on their difficult quest. Marcy chose subjects in particle physics and cosmology – difficult, highly theoretical material.

But his love of astronomy remained. He recalls a lecture given by Carl Sagan during a visit to Los Angeles during Marcy's second year of study: 'His message was wonderful. Carl combined astronomy with a vision of a Milky Way teeming with extraterrestrial life, which we will one day be able to link up with. It was technological, but also ideological, a kind of harmonious society. That was when the extraterrestrial really came alive for me, as it did for a lot of others.'

Marcy graduated with first-class honours in physics and astronomy. Career opportunities were there for the taking. He was offered PhD places at the prestigious universities of Berkeley and Caltech, the California Institute of Technology. But he chose the University of California in Santa Cruz, not far from San Francisco, which ran the Lick Observatory – the observatory where Bill Borucki would later build his Vulcan Telescope. Students at Santa Cruz were allowed to use the biggest, 3-metre telescope at Lick. 'I already knew then that I didn't want to become a theoretical scientist,' Marcy tells me. 'I was a telescope guy.'

After attending lectures for a couple of years, it was time for Marcy to choose a subject for his dissertation. 'I didn't know what to do. I'd always thought I would become a cosmologist and study the origins of the universe, or maybe the fundamental forces of nature . . . I wanted to study everything. It was so silly, you know,' he says, shaking his head. He was terribly unsure of himself. 'I'd known from high school that I was only of average intelligence. I remember sitting in the biology or physics class and not understanding a word the teacher said. There were always a few other pupils who did get it and got the best grades.'

Marcy speaks very openly about his feelings of inferiority. He tells me at least eight times during our conversation that he was by no means the smartest kid in the class. It's quite strange. I'm sitting in the office of one of the most renowned professors at one of the top universities in the u.s., whose name appears on the list of Nobel Prize candidates – and he continually tells me about his limited intelligence. I counter by saying that he must have done something good to have been awarded so many research grants. Wasn't his mediocrity just relative? 'Yes, I guess so,' he admits. 'But I'm really not that good at solving equations. And that is what gets tested at high school and university. That is actually a great pity. I didn't discover my strong points until much, much later. My talent is for asking the right question. A profound and interesting question that you can immediately come up with an experiment for to answer a little part of it.'

At Santa Cruz, he had not yet discovered these talents. George Herbig, a professor at the university, took him under his wing. Herbig was one of the first to map out the regions where stars are born and describe the process of star formation. He did that by studying the spectrum of starlight.

The white light emitted by stars is made up of many colours. Projecting a ray of starlight through a small opening will cause it to fan out. Its spectrum can be seen as a rainbow on a detector, which then photographs it. This technique, known as spectroscopy, tells us a lot about the star emitting the light, including its velocity, its temperature and its chemical composition. Marcy would later make good use of spectroscopy. He also learned a lot about his mentor's approach to science. 'Herbig interpreted observational data in a minimalist way,' he tells me. 'He only described the observed data and embellished them with as few unnecessary details as possible. That makes you feel sure of yourself, and you don't make as many mistakes. I've worked that way throughout my career.'

Marcy was awarded his doctorate in 1982 and secured a good grant to continue his research. He used spectroscopy to observe the magnetic properties of stars. He was given access to the Mount Wilson telescope, where Edwin Hubble had famously measured the expansion of the universe in the late 1920s. On paper, Marcy had a dream career, but he felt more insecure than ever. 'I saw no future in my research,' he tells me. 'A Harvard professor had been very critical of my dissertation. Everyone around me was successful, and I felt like an imposter. By accident, I found the reference letter that Herbig had written about me and it was pretty lukewarm. That sort of thing discouraged me. But I didn't want to stop either. What would my friends say? None of them knew me well enough to know that I actually wasn't that smart. What would my parents say? And especially my father. I didn't want him to see me as a failure.' Marcy went to see a therapist to help him ignore the criticisms and the pressure to compete. But it didn't solve the underlying problem. Marcy had ended up in the wrong area of research. It was too complicated and it didn't interest him enough. He wanted to find something that he felt was a calling, something that would spark his enthusiasm.

The story of the shower that Marcy took one spring morning in 1983 has been told many times over, down to the last detail. Google 'Geoffrey Marcy shower' and you will find several versions of this life-changing event. It may be exaggerated here and there, but when he describes it to me, he is still full of passion, without sounding implausible. On the contrary – the bathroom is well-known as a place where inspiration can suddenly appear from nowhere. Wasn't Archimedes taking a bath when he had his renowned Eureka moment? 'I was standing in the shower and thinking, what should I do, what should I do? I must have been there for three-quarters of an hour. The water poured over my back. It felt good, and I was able to let my thoughts run free. And I thought, okay, if it's going to fail, then let it fail honourably, or at least poetically. I can use the big telescope for another year. So I'll use it to do something unique, something that has a special meaning for me. Not to please other astronomers, not to impress my friends, not to make my father happy. Just for me. The question I asked myself was, are there planets around other stars? I thought about what I'd learned while I was studying. A star that has a planet makes small movements because of that planet. I could use spectroscopy to measure velocities. Perhaps I could measure those movements very accurately. Then maybe I could find objects that are smaller than any that had been found so far. Maybe I could find planets.'

A dejected researcher had stepped into the shower. When he came out again, he was a man with a mission. The boys' storybook scenario was complete.

MARCY WANTED to measure the same effect as Captain Jacob, Peter van de Kamp and the other astrometrists – the wobble of a star caused by a planet. Planets originate in the same rotating cloud of dust and gas from which stars are formed. A planet continues to circle a star because it is held in its orbit by the star's gravitational pull. But the star is also slightly affected by the planet's gravity, which makes it 'wobble' a little. In effect, the planet and the star rotate around each other. The existence of the planet can be confirmed by measuring the star's wobble. Once the velocity and orbital period have been measured, Kepler's laws can be applied to calculate the planet's mass and the radius of its orbit.

Marcy wanted to measure stellar wobble using the Doppler effect, most famously explained by the passing of an ambulance. The tone of the siren becomes higher as the ambulance approaches and then lower again as it gets further away. That is because sound is made up of waves. As the source of the sound comes closer, the waves are compressed and therefore become shorter, resulting in a higher tone. As the source becomes more distant, the sound waves are elongated again and the tone sounds lower.

This effect was predicted in 1842 by Austrian physicist Christian Doppler and first proved a few years later by Dutch meteorologist and founder of the Royal Netherlands Meteorological Institute Christophorus Buys Ballot. He was given permission to use a locomotive on the newly laid railway line from Utrecht to Maarssen. He posted professional horn players at fixed points along the route, who would play a set note. From the locomotive, other musicians – 'whose hearing was of such a degree of accuracy and refinement that it may be equalled but was difficult to attain' – listened to ascertain the change in tone. After a first attempt in the winter of 1844, which was abandoned due to 'a sudden storm of hail and driven snow', the experiment succeeded in June of the same year. The musicians on the train clearly observed that the note produced by the horn-players alongside the line became lower in tone as they passed and moved further away.

The same trick can be applied to moving stars, as their light also consists of waves. Blue light has shorter waves than red light. A star that is moving towards the observer should then look a little bluer, while one that is moving away will look redder. Doppler had already predicted that 'this will in the not too distant future offer astronomers a welcome means to determine the movements and distances of . . . stars which . . . until this moment hardly presented the hope of such measurements and determinations.'

The Doppler colour shift does indeed exist, but is very, very subtle. The minuscule changes in colour cannot be observed with the naked eye, but can only be seen by looking at the shift in the absorption lines in the spectrum. If you look at a star through a prism or a spectrograph and project the spectrum on a screen, you will see that the rainbow of colours is interrupted in certain places, like a sort of barcode. Every stripe has a certain wavelength where the star is darker; this is because part of the light from its interior is absorbed by the outer layers of the star's atmosphere. The wavelengths of these 'absorption lines' are determined by the chemicals in the star's atmosphere. Hydrogen, helium, iron and oxygen all have their own 'barcode' or signature of absorption lines. If a star moves, the Doppler effect will cause the colour – and thus the wavelength – of the absorption lines to shift slightly. A line that is normally found in the blue part of the spectrum will move a little towards the red side of the rainbow. This shift can be used to calculate the velocity of the star. Because the star and the planet move in circles, their velocities will continually change. The regularity of the period in which this happens can be used, by applying Kepler's laws, to determine the mass of the planet.

It was not this measuring method that made Marcy's plan original. Astronomers had been using spectroscopy and the Doppler shift to measure the speed of stars for nearly two hundred years and it had led to the discovery of many binary stars. Early planet hunter Otto Struve had already predicted in his prophetic article in 1952 that the method could also be used to discover exoplanets. Marcy's plan was above all original in that it was doomed to failure, for two reasons: accuracy and time.

LIKE BILL Borucki, who was at that very moment busy refining the transit method just a few dozen kilometres away, Marcy asked much too much of the available technology. And he, too, was engaged in a lonely quest. 'The smallest velocity that you could measure with the spectrometers of the time was around one kilometre per second,' he tells me. 'We wanted to measure speeds that were a thousand times smaller. What's more, I had to be careful what I said to my colleagues. Hunting planets was ridiculous. Almost embarrassing. If you told people you were hunting for planets, it was as though you claimed to have discovered a paranormal source of energy in the pyramids of Giza. They would choke on their ham sandwich, look at their shoes or start talking about the baseball results.'

The time factor made the search even more hopeless. Observing the movement of a star was not enough on its own. To prove that the movement was caused by a planet, Marcy had to see the star move *back and forth*. And to be really sure that it was not a random movement, he had to see the planet orbit the star at least three times. How long would he have to wait to see that?

Marcy knew just as little about what a star's planetary system looked like as the Order of the Dolphin and the pioneers of the Kepler team. The only example available was our own solar system: four small inner planets with orbital periods of around a year, and four large outer planets with orbits of tens to hundreds of years. With his method, he could only focus on the most massive planets in a system, because they had the strongest gravity. In a duplicate system of the solar system, he would only be able to measure the influence of a Jupiter. Jupiter is an outer planet that takes almost twelve years to go around the Sun. To determine the existence of a planet with certainty, Marcy had to observe three full orbits. That would take 36 years. Marcy knew that if he were ever to discover the first exoplanet, it would not be until after he had retired. He puts on a charming smile. 'I knew that no one would believe me until I had observed three orbits. We needed a great deal of luck to make the discovery, and then we would have to wait a very long time to confirm it. I was always in it for the long haul.'

FORTUNATELY, MARCY had a plan B, a way out for both the accuracy and the time problems. In the mid-1980s the small 'almost stars' that Jill Tarter called 'brown dwarfs' were very popular. Stars only account for a very small fraction of the total mass in the universe. There must therefore be a large, hidden quantity of matter that does not radiate light. This is known today as 'dark matter' and is still an unsolved mystery. Brown dwarfs, which are ten times less massive than the Sun and radiate almost no light, were considered potential candidates for this hidden mass. So people wanted to know how many of them there were – a legitimate research question, totally free of 'little green men'.

'Our search would identify brown dwarfs as well as planets,' Marcy says. 'A brown dwarf is fifty times more massive than Jupiter. If it is in close orbit around a star, the star will make small circular movements (wobbles) with a speed of up to 1 kilometre a second. That could easily be observed with the spectrographs of the time. Because of the planet's short orbital period, the wobble would also last less than a year, so that research results could be produced more quickly.' Besides a sound motive for conducting research, the brown dwarfs also gave Marcy a cover for his real purpose: searching for planets. 'These days, everyone sticks the word "planet" in their article or funding proposal to make it interesting,' he tells me. 'But back then, I actually avoided it in texts that would be assessed by others. I always used the term "sub-stellar objects", leaving it in the middle whether I meant brown dwarfs or planets. That was less controversial.'

Marcy obtained a position as a professor at San Francisco State University. Alongside his busy teaching programme, he kept a little time free for research. His aim was to improve the existing spectrographs so that they could measure smaller speeds and therefore detect smaller objects. The great problem with spectrographic measurements of the Doppler shift (and thus of the velocities of stars) is the reference spectrum. You want to see whether a star's absorption line has shifted due to the Doppler effect – that is, has become a little bluer or redder – compared to its wavelength 'at rest', that is, the wavelength of a stationary source in a laboratory. To do that you need a reference spectrum: the 'signature' of absorption lines of a gas that is not moving. You can, of course, measure the reference spectrum in the laboratory beforehand and then compare it with the star's spectrum. But the Doppler shifts are so minimal that they can hardly be distinguished from a small error in measurement. It is easy for such an error to creep in between the two measurements – for example, because a lamp is not correctly aligned or a screen moves half a millimetre.

Marcy devised the trick of measuring the star and the reference spectrum *at the same time*. He wanted to shine the light from the star through a reference gas so that he could record the shifted spectrum of the star together with the stationary reference spectrum. This method made errors in measurement almost impossible. If the measuring equipment displayed irregular oscillations, the stellar and the reference spectra would be affected to the same extent. The differences between their wavelengths would stay the same – and that is what he wanted to measure.

At a brown dwarf conference, Marcy spoke to Bruce Campbell, a Canadian astronomer who, together with his colleague Gordon Walker, was leading a large-scale research project into sub-stellar objects. Campbell was using the new spectroscopic method, with hydrogen fluoride as a reference gas. It worked well, but Marcy wanted an alternative for the lethal hydrogen fluoride. 'That stuff eats your lungs up while you don't even know it,' he says cheerfully.

Marcy finally decided to use iodine. It was a lot less dangerous and was excellent as a reference gas. Together with student

Paul Butler, Marcy designed a glass casing for the iodine in the spectrograph, which had been devised by Steven Vogt, one of his mentors at Santa Cruz. Marcy takes a small model of the glass tube from his bookcase, where it had been standing next to a book entitled *Iodine: Why You Need It, Why You Can't Live Without It*.

Together with Butler, he developed a computer program that could distinguish between the stellar and the reference spectra. It took them six years to write the software, by which time the spectrograph at Lick Observatory was due an update. Once that was out of the way, Marcy and his students could finally get down to making observations. It was a colourful team, with a few students who did not exactly fit the standard academic profile. Mario Savio was a civil rights activist and was an icon of the American peace movement in the 1960s. He became interested in planetary research because it would push back social frontiers. Another student, Debra Fischer, had worked as a nurse for many years before studying physics and later joining Marcy's group. She now leads an important exoplanet research group at Yale University.

Marcy digs out a thick, well-used book, which is still on his bookcase after all these years. It contains long lists of stars. More than a hundred of them are marked with a cross. These are stars that were considered eligible for having planetary systems because they were bright enough and were in the same stage of their lives as the Sun. Marcy and his colleagues observed them regularly through the Lick telescope. The measuring equipment and the software worked. But the team found no planets. Marcy was not worried – after all, it could take years for the data to reveal the presence of a Jupiter-like planet. And it would have to make a series of orbits before being confirmed. The team's motto was to gather and process, gather and process, gather and process. Until that autumn day in 1995, when Marcy received the message from Italy about the star 51 Pegasi. He shows me the name on the yellowed page of the star catalogue. There is no cross in front of it.

'WE HAVE no choice,' Michel Mayor told his son. 'I think we're going to have to take you down a peg or two.' During a holiday in Provence, Mayor and his friend and colleague Didier Queloz used code names for 'that one star'. They talked about it a lot around the dinner table. The two families had redubbed the observation trip a holiday; the Observatoire de Haute-Provence was beautifully situated. Two white observatory domes reached out above the treetops in the undulating landscape. A short distance away were the sandstone houses of the village of Saint-Michel-l'Observatoire with their red roofs. The men spent their nights at the observatory. And now Mayor's son had finally discovered the name of the mysterious star that had held his father in its grip for the whole holiday: 51 Pegasi.

Fortunately, nothing came of the plan to have the boy quietly disappear. Eighteen years later he is also a scientist, as are his two sisters. Their father Michel smiles at me from my computer monitor. The image of his friendly face, with a grey beard and glasses, is a little shaky on Skype. Mayor is an emeritus professor at the University of Geneva, where he has worked for almost fifty years. He started as a doctoral student and worked his way up to become director of the observatory. His retirement does not mean that he has stopped working. In the past year alone, he had co-authored more than twenty publications. The day after our conversation, he left for Vietnam to give an astronomy course to young people. While researching his PhD in the late 1960s, he developed theories about how the stars in the Milky Way move. He wanted to test those theories by measuring the velocities of stars. With the Doppler effect.

'When we first started, no one was interested in measuring the speed of stars,' he tells me. 'You had to measure the tiny

Doppler shifts in the spectrum by hand using the detector screen. But you could never achieve a decent degree of accuracy manually. It was an extremely dull and unsatisfying activity.' At the end of the 1960s, he heard of a method that could be used to measure Doppler shifts much more quickly and accurately. The idea for the method had been formulated in 1953 by light expert Peter Fellgett, who said 'If the only aim is to measure the Doppler shifts, it's not necessary to observe all the details of the stellar spectrum.'

The secret was to use a special accessory on the spectrograph, a kind of template. It consisted of a glass plate coated with metal placed in front of the screen on which the stellar spectrum was projected. There were holes in the plate where absorption lines were expected. If the absorption lines had shifted, you had to move the plate to one side a little until the lines were completely aligned with the holes. That meant that you lost the rest of the information on the spectrum, but that was no problem as all Mayor wanted to measure was the speed of the star. By measuring the change in position of the plate, Mayor could immediately see the Doppler shift and thus the speed of the star. It was simple and more effective than the method used by Marcy and Butler. While Paul Butler had to first analyse his spectra by sitting at the computer for several hours, Mayor could read the speed directly from his measuring instrument. The measuring speed would eventually prove decisive in the race to discover the first exoplanet.

MAYOR DID not initially focus on exoplanets. With his first version of the spectrograph, he worked with his colleague Antoine Duquennoy on mapping binary stars close to the Sun. The research lasted for more than fifteen years and was completed in 1991. 'Measuring velocities was basic astronomy,' Mayor tells me. 'But it had never been done with so many stars at the same time.' It led to his second most cited article, which was quoted no fewer than 1,976 times in 2014, 35 times more than his most famous article, from 1995, in which he presented the discovery of the first exoplanet.

One of the stars in his large-scale observation programme, HD 114762, displayed strange movements, suggesting that there might be a brown dwarf, or even a planet, orbiting it. Mayor co-authored a *Nature* article about this object. Harvard professor David Latham managed with great difficulty to forestall the media circus that threatened to erupt as a result of the discovery of this 'maybe planet'. To this day, no one has succeeded in proving whether the object is a small brown dwarf or an overgrown giant planet.

In 1990, partly as a consequence of the article, Mayor found himself at a conference on 'bioastronomy'. The location, the village of Val-Cenis in the French Alps, was close to home. It had perhaps been chosen because of the name of a nearby mountain, Mount Seti, which the conference attendees visited. Besides co-organizer Jill Tarter and SETI veterans like John Billingham and Frank Drake, a broad selection of scientists from various disciplines were invited. It was a sequel to the Green Bank conference of 1961, where the Order of the Dolphin had been born. As back then, the topics of discussion were based on the various terms in the unsolvable Drake Equation. Questions were asked on astronomy (for example, 'The Habitability of Mars-like Planets around Sun-like Stars'), biology ('Self-assembly Properties of Primitive Organic Compounds') and intelligence ('Cognition in an African Grey Parrot'). Some were a little far-fetched (such as 'Biological Constraints on Interstellar Travel' and 'The Likely Organizational Order of Advanced Intelligences'). After the conference, the search for exoplanets was high on Mayor's research agenda.

Tragically, in 1994 Antoine Duquennoy was killed in a car crash. By that time, a new spectrograph designed by Mayor and André Baranne, an instrument-builder from Marseille, had been mounted on the telescope in the Haute Provence. This new spectrograph had a greater degree of precision and could detect minute differences in speed, enabling it to identify binary stars with small companions, which could be as little as a planet. They compiled a list of bright stars, to explore which of them had such a companion. One of the 142 stars on the list was 51 Pegasi. 'Marcy and Campbell's teams had searched for years and not found any planets,' Mayor tells me. 'So we deliberately selected stars that were not on their list to increase our chances of a scoop. The reason that Marcy had not chosen this star – that it was of a different chemical composition than the Sun – was not important to me.' Because it was so bright, 51 Pegasi – which can be seen with the naked eye – was used as a control, one of the handful of stars that they observed especially often. They watched it almost every night, as an extra test for the stability of the spectrograph.

Twenty years after starting his doctoral research, Didier Queloz looks back on the events of the winter of 1994. He is now a professor at Cambridge and is visiting Amsterdam to give a lecture. I talk to him over lunch in the department's favourite local bar. As he attacks a prawn croquette, he confides in me: 'One of the factors that later proved conclusive in the race with the Americans was that Michel Mayor was not there that winter. He was spending the winter in Hawaii and let me do the first observations for our programme.' Towards the end of 1994, after observing the star for several weeks, Queloz noticed that there was something strange about 51 Pegasi: it was moving back and forth. On one evening it was moving away from the Earth at 40 metres a second, while on another it was getting closer at the same speed. 'I could have concluded at that moment that the spectrograph was not working, or switched to another control star. But because Michel was not there, I was free to decide for myself. I became obsessed by the star.' Queloz increased the number of observations to find out what was going on. The changes in speed seemed regular. The star moved gradually back and forth with a period of 4.2 days and a maximum speed of 60 metres per second. Together, the speed and period gave the planet's mass and orbit. The motion had to be caused by an object about half the mass of Jupiter and with a very small orbit: an exoplanet.

In February 1995, Queloz sent a fax to Mayor in Hawaii to tell him about the star's movements. 'I said that it could be a planet,' he tells me. 'Michel gave the best answer I could imagine: "Perhaps." Most PhD supervisors would have dismissed such a suggestion outright as nonsense.' The short orbital period was completely unexpected. Kepler's law dictates that the closer a planet is to a star, the shorter its orbital period. Jupiter takes nearly twelve years to go around the Sun, the Earth a year and Mercury only 88 days. A period of 4.2 days meant that the planet was only 8 million kilometres from the star, less than a twentieth of the distance between the Earth and the Sun. The planet of 51 Pegasi was half the size of Jupiter and eight times closer to the star than Mercury to the Sun. In the edible model of the solar system I had built in the school classroom, the exoplanet would be a pea only 78 centimetres from the star, a large red cabbage.

As everyone knows, such a planet could not exist. Planets are remnants of a star's formation process. Stars are created from dust and gas clouds, which collapse under their own gravity. The material in the cloud then collects in a flat, rotating disc. The dust and gas spiral slowly towards the centre where, over a period of a few million years, a star grows. Eventually the disc disappears, leaving a few clumps behind – the planets. A number of these discs were discovered in the 1970s and '80s, supporting this theory of planetary formation.

In January 1995, while Queloz's observation programme was still in full swing, authoritative professor Alan Boss wrote in *Science* that a large quantity of cold matter was required to form large planets. It was so hot close to stars that only small planets, like the Earth, could be formed. Boss stated resolutely that Jupiter-like planets should form at distances of approximately four to five times the distance from the Earth to the Sun. Jupiter itself is a little over five times further away from the Sun than the Earth. Mayor and Queloz realized that, according to Boss's predictions, their planet could never have formed where it was now in orbit around 51 Pegasi.

After February 1995, the star could no longer be observed. The night sky changes with the seasons and, in the spring, 51 Pegasi was in the sky during the day rather than at night. Mayor and Queloz decided to postpone publishing their findings until the summer, when it would be possible to see the star again at night. They would then be able to observe its movements once more, just to be on the safe side. Before announcing such a controversial discovery to the world, they had to be absolutely sure of themselves and that they were not on the wrong track, like See and Van de Kamp. And, besides, an extra week in Haute-Provence was a good excuse to take a holiday.

In July, their earlier measurements were confirmed. The star 51 Pegasi still moved back and forth in the same period and at the same speed. There was little doubt that they had discovered the first exoplanet around a Sun-like star. At the dinner table in their holiday home, where Mayor's curious son joined in the conversation, they finally decided to write up their results and send them to *Nature*. The families celebrated the discovery with cake and champagne.

On 29 August 1995, Mayor and Queloz submitted their article. As they wanted to exclude all possible errors in their interpretation and all alternative explanations for their observations, they asked friends for advice. Wouldn't such a massive planet evaporate so close to the star? Wasn't it a brown dwarf? Were the fluctuations in the star's speed real? They left essential details out of their correspondence, but their colleagues sensed that it was something important. 'I had registered us for the Cool Stars conference in Florence in October 1995,' Mayor tells me.

'More than 300 experts would be attending. It was the perfect moment to announce our discovery. I was too late to register for a lecture, but they could give me five minutes speaking time. That was very little, but I'd purposely been vague about what I was going to say.'

Somehow, the news of the announcement leaked out and the rumour mill started spinning at top speed. Mayor received a call from a colleague who was to give a lecture at another astronomy conference in September in Sicily. He offered to promote what Mayor was going to say in Florence; Mayor agreed. 'He announced that, in October, we would present the discovery of the first exoplanet,' he tells me. 'After that, I was informed by the organization in Florence that my speaking time had been extended to 45 minutes.'

When Mayor and Queloz travelled to Italy, *Nature* had not yet accepted their article. It had been sent to three different colleagues, who had to assess independently whether it was innovative and soundly enough argued for publication in this prominent journal. One of them was Alan Boss, who had calculated earlier that year that Jupiter-like planets could not be formed so close to their stars. Another colleague, observation expert Gordon Walker, had expressed doubts about the methods they had used. In short, the jury was still out when Mayor and Queloz arrived in Florence.

FLORENCE IS the city of Galileo Galilei, who discovered the moons of Jupiter four hundred years ago.[19] In the first week of October 1995, celestial bodies circling each other were once again discussed in Florence. And this time, too, the subject-matter was controversial: the discovery of the first planet around another star was to be announced. Although there were plenty of critics at the conference, the discoverers – unlike their Italian predecessors – were not threatened with torture.

'Everything happened at once. It was insane. O. J. Simpson was acquitted that week, the Boston Red Sox were in the play-offs, and then we had Mayor and Queloz's discovery.' Andrea Dupree, researcher at the world's largest astronomical institute, the Harvard-Smithsonian Center for Astrophysics in Cambridge, Massachusetts, remembers Cool Stars in Florence very clearly. She was one of the organizers. 'Mayor and Queloz's discovery was greeted with applause, but a lot of people were still sceptical,' she tells me. 'And then there was all that media. Poor Michel.' She plays 'The Planet Hunting Song' to me, which was written a few years later at the next Cool Stars conference. She hums quietly along to the tune. Fortunately, just as I'm wondering if I ought to join in, she presses the stop button and says 'Boy, I haven't heard that song in a *looong* time. Anyway, you get the idea.'

One of the other organizers was Francesco Palla, now former director of the Osservatorio di Arcetri in Florence. I speak to him in his office at the magnificent old observatory. His window looks out across the green hills of Tuscany and Galileo's old house, hiding beneath long cypress trees. It was there that he wrote his *Dialogo*, the book in which his heliocentric world-view triumphed over geocentrism. Galileo spent his final years there, blind and in exile.

Palla is nearly sixty, but looks much younger.[20] He has a tanned face, thick salt-and-pepper hair, round glasses and a fashionable thin pullover over a well-ironed shirt. Photographs of Renaissance paintings and nebulae hang on the walls of his office and there are a few physics instruments on his bookshelves. He shows me pictures of the closing dinner of the conference. The diners at the round table wear 1990s glasses and corduroy trousers. Among them is Michel Mayor, his hair still black and wearing a kind of Hawaiian shirt. 'Nice shirt for a formal dinner,' Palla says in mock-disapproval.

'This one was taken on 6 October, the day after the dinner, when Mayor announced his discovery,' Palla says. 'Look, he wore a tie for the occasion.' We see Mayor in a bare lecture room with no windows and a low ceiling. The only indication that it is in Italy is the bottle of wine on the table behind him. The bearded astronomer, in socks and sandals, is using a long wooden stick to point to the wall, on which an image has been projected. It is the graph – later to become famous – showing the back and forth movement of the star 51 Pegasi. The audience in the front row are listening attentively and making notes. We leaf through the photos, but do not find a better picture of this important moment. 'The photographer was a nice guy,' Palla says. 'But he made a real mess of the pictures. Look, all you can see is people's backs.'

The journalists sat in the back row. They had come from all over Europe and even the *Washington Post* had sent a reporter. A *Rai Uno* television crew, invited by Palla, was making an item for the evening news. One of the people avidly taking notes in the front row was Harvard professor Bob Noyes. There was a good reason for that, as Noyes explains to me in a long email. There were rumours at the dinner that Mayor would be revealing important news the following day. When Noyes asked Mayor about it, Mayor told him in confidence that he had discovered the first exoplanet, with an unexpectedly short orbital period of 4.2 days. He also let slip the name 51 Pegasi.

The morning after, Noyes had called his colleague Tim Brown at the Mount Hopkins observatory in Arizona, where it was still night-time. Brown was busy with their own programme, searching for planets using the Doppler method. He picked up the phone with a cheerful '*Buon giorno*, what's up?' and reacted with surprise when he heard the name of the star: '51 Pegasi? I was looking at that just five minutes ago!' He immediately planned a series of extra observations, to confirm the short orbital period in the next few nights. That same evening, after the announcement, another conference attendee tipped off Geoffrey Marcy, who was making observations at another observatory in the U.S..

Palla apologizes that there are no minutes of the meeting. 'It was chaos,' he says. 'Typically Italian.' Questions and answers followed each other in quick succession. There were experts from a wide range of disciplines in the conference room, but no one had ever heard of a planet around another star. Most of the questions were about the planet's location in relation to its star. Alan Boss had after all calculated earlier that same year that Jupiter-like planets could not be formed that close to a star. The only solution they could come up with was planetary migration, in other words, the planet had formed at a greater distance and had later moved closer to the star. No one knew exactly how that worked. They cautiously accepted that the planet was there. But how it got to be there remained a mystery.

As far as the press was concerned, Mayor was in a difficult position. While the article was still under consideration by *Nature*'s panel of experts, he was not permitted to say anything about it. He could not talk to journalists about his discovery, but he was allowed to present it to other astronomers. And the journalists were in the same room as the scientists. 'Afterwards, journalists were fighting among themselves to get a personal interview,' he tells me. 'When I got back to the hotel late in the evening, there was a gigantic pile of faxes from newspapers and press agencies asking for comments. It was a weird situation – everyone at the conference was free to talk about it except us.'

Many of the headlines were sensational: 'BY JUPITER! ASTRONOMER FINDS SUN'S SISTER!' Others were speculative: 'MYSTERIOUS PLANET KEY TO EXTRATERRESTRIAL LIFE?' The articles mainly cited sceptical colleagues of Mayor and Queloz. Even the local Dutch newspaper *Eindhovens Dagblad* jumped on the bandwagon, quoting an astronomer from the University of Leiden who did not believe in the exoplanet as saying 'A Jupiter-like gas giant that close to a star would have evaporated long ago.' The *Washington Post* mistakenly reported that the discoverers were Italian, rather than Swiss. None of the articles reported the first incontrovertible observation of a brown dwarf, a major discovery which had also been announced at the conference. The planet in Pegasus demanded all the attention.

'Didier and I said to each other that we would simply wait until we could speak freely,' Mayor says. 'Then we would talk to the press for a while and, eventually, all the attention would die down. But that never happened. It is now twenty years later and public interest has never faded. Film crews and journalists are always ready to hear about the discovery of new planets. Always, always, always.'

IN A book that he wrote later, Mayor introduced Geoffrey Marcy as 'Dr Death', a nickname he earned because of his ruthless rejection of competitors' claims to have discovered planets. 'Notwithstanding his amiable personality,' Mayor added, by way of conciliation. At first, Marcy did indeed respond to the discovery of the 51 Pegasi planet by saying 'Once again a hope which I'll be obliged to dash.' But after his four nights on Mount Wilson he had seen the star's movements with his own eyes. 'I was dumbstruck,' he says. 'First. Exoplanet. Ever.' He wrote an enthusiastic email to Mayor, saying:

> So your wonderful discovery is confirmed!!! Congratulations again! *Alors*, I will be asked by journals to comment on 51 Peg. Please tell me if there is any problem with my stating to them that we confirm the observational result.

Tim Brown in Arizona also confirmed the planet's four-day orbital period. Mayor, Queloz and Marcy and Brown's groups sent a joint telegram to the astronomical community, reporting that the three groups had – independently of each other – observed the four-day orbit around 51 Pegasi. The Americans were not affected by *Nature*'s restrictions and, especially Marcy, spoke freely to journalists. American papers reported that their compatriots, too, agreed with the claims from Europe. That helped increase the credibility of the discovery, in any case among the wider public. At the end of October, Mayor and Queloz received notification from *Nature* that their article had been accepted.

The question of how it had come to be so close to its star remained unanswered, but the existence of the planet was indisputable. It was given the official name 51 Pegasi b – exoplanets are named after their star, suffixed by a small letter. It starts with b, and any other planets found around a star are given successive letters. Creative names like Tatooine and Cybertron are reserved for planets in films and comic strips.[21]

Mayor is laconic when I ask him if Marcy's press campaign may have helped get his article accepted by *Nature*. In theory, Marcy and Butler – or Noyes and Brown – could have published their own independent observations more quickly, thereby staking their claim to have discovered the first exoplanet. Not a single planet in our own solar system had been discovered by an American (except Pluto, but that had always been the odd man out). How symbolic and appropriate would they have found it on that side of the Atlantic if the first planet outside the solar system had been a 'final frontier' explored by the u.s. But Mayor was never afraid that Marcy would claim the discovery for himself. 'We'd already presented it in Florence,' he says. 'Although the article had not yet passed the assessment by *Nature*, everyone knew that we were the first.' Even if Marcy had wanted it otherwise, he had no other choice than to talk of a Swiss discovery in all his interviews.

After the discovery of 51 Pegasi b, Marcy and Butler examined all the observation data they had collected on other stars. Until then, they had only looked for orbital periods of several years. Now they conducted a calculation to see if there were other shorter periods like that of 51 Pegasi. And there were. Without having to make a single extra observation, they found two other planets in their existing data. Like 51 Pegasi b, both were giant planets close to their stars, with periods of only a few days. The existence of these 'hot Jupiters' (or, in the words of some light-hearted press reports, 'barbecue planets') had been predicted by Otto Struve. He had said in 1952 that there was 'no compelling reason why the hypothetical stellar planets should not, in some instances, be much closer to their parent stars than is the case in our solar system'. The discovery of these hot gas planets was a surprise to most astronomers, who had assumed that other solar systems would resemble our own. The existence of hot Jupiters was not considered impossible; it was simply completely unexpected. The first exoplanet had been right in front of Marcy and Butler's noses for many years, but they hadn't seen it.

Marcy smiles broadly when I ask him why, despite missing out on discovering the first exoplanet, he was so delighted for Mayor and Queloz. Why was he so euphoric when he drove down the mountain on the morning of 10 October 1995? 'I had been looking for exoplanets for so long, without success,' he explains. 'No one thought they existed. I was so happy that the search had not been for nothing. I felt like I was on Columbus's ship.'

Even though he may have been on the second ship in Columbus's fleet, Marcy revealed the two new planets with a great display of drama. At the beginning of 1996, he presented the discovery at a press conference of the American Astronomical Society. Scientific journalist Michael Lemonick was there and later recalled that Marcy 'spoke with the theatrical flair that I would later recognize to be his trademark. "After the discovery of 51 Pegasi b," he said, "everyone wondered if it was a one-in-a-million discovery. The answer is . . . no. Planets are not rare after all."'

For Marcy, this was a positive end to a bittersweet episode. To be the first to discover a new kind of celestial body is, of course, the dream of every astronomer, but such a discovery is only of limited value if no more similar objects are found. One swallow doesn't make a summer. The Swiss had discovered the first planet, but the Americans had numbers two and three and were once again ahead in what was to become a great planet race. Since then, both groups have found hundreds of exoplanets using the Doppler method. The Swiss team is still miles ahead of the Americans, though it is no longer really a two-horse race: both teams now consist of many nationalities, the American group has divided into smaller factions, and sometimes they even work together among the ever-growing number of planet-hunting research groups. In the rumour mill, the three names that are inseparably linked to the discovery of the first exoplanet feature regularly on lists of candidates for a Nobel Prize: Mayor, Queloz and Marcy.

Mayor and his Swiss colleagues laugh and shrug their shoulders when I tell them that, at universities in the u.s., I have largely heard about American discoveries. The Geneva team is undeniably way ahead in the development of the Doppler method. And they have placed the third upgrade of their spectrograph on the 8-metre-wide Very Large Telescope – vlt for short – in Chile.

Marcy, who retired from the Berkeley faculty a few years after I met him,[22] previously used the equally gigantic Keck Telescope (which is 1 metre wider than the vlt) in Hawaii for his research. Paul Butler, who was always in the shadow of his flamboyant colleague at their presentations, has not worked with Marcy since 2007. In astronomical circles, the break between the two was referred to as 'the divorce'. Their 'worldly goods' – the observation time on their new shared telescope – was divided fairly between them.

For years, the mediagenic Marcy was one of the world's most prominent planet hunters. He popped up everywhere, even on David Letterman's talk show. A researcher at Berkeley told me that he was once in a bar talking to colleagues about the media frenzy surrounding Marcy when the professor's face suddenly appeared on the television screen above their heads. At the next

Christmas party, they made a short film in which Marcy could read and control everyone's minds, like a kind of Big Brother. Marcy himself, of course, thought it was hilarious. It is difficult to believe that he has never been upset about his failed scoop. But it is equally difficult to imagine him gnashing his teeth about it.

seven

THE HUT IN THE CAR PARK

THEY OPENED the roof and looked out at the spectacular night sky. There was hardly any room in the tiny hut. The tallest of the two men had to bend over until the roof was open. It was a cool September night in the Rocky Mountains. They had put on their thick coats and there was a flask of coffee on a small table pushed into a corner. The car park was almost empty. The lights of the institute were dimmed; on the other side of the hut, where the lake was, it was pitch black. The conditions were good for observation. The men set their telescope to the coordinates they had received from Boston. For ten hours, they photographed the star every minute. It would be two months before they could look at the results of their observation on that night of 9 September 1999.

Four years had passed since the discovery of 51 Pegasi b. In the meantime, some ten planets had been discovered. They were, in their majority, hot Jupiters – big, massive and close to their stars. The strong gravitational pull of the planet made the star wobble enough to be observed with a spectrograph. In a few cases, stars had been discovered with several planets in orbit around them. And yet, there were still many other interpretations of the variable Doppler shift of these stars. It could be caused by objects more massive than planets, or pulsations of the star itself. Exoplanets would not be accepted by the astronomical community until their existence had been confirmed by another, independent method. The two researchers in the far too small hut wanted to see a transit – an exoplanet moving in front of a star. This method had already been predicted and worked out by many others, including Otto Struve, Frank Rosenblatt and Bill Borucki, but had never been successfully put into practice.

THE YOUNGEST of the two researchers, who had just started his doctoral research in 1999, is now a professor at Harvard. I meet him one day in the autumn of 2013 at Leiden University Observatory, where he is visiting to give a lecture. David Charbonneau is approaching forty, a tall man with blue eyes and black hair parted to the side. He wears a smart jacket and jeans. He speaks in eloquent, subtle sentences; not as lyrical and impassioned as Geoffrey Marcy, but he clearly enjoys explaining his subject, not only to his students but to the wider public. In the u.s., Charbonneau is an ambassador for planet hunters and is a frequent guest on talk shows and current affairs programmes.

'I have to be a little careful,' he answers when I ask him why he chose astronomy. 'A lot of people tend to rewrite their own history.' He may be right. Most astronomers that I interview have their stories off pat. Walks on the beach with their fathers, their first telescope, pondering on their futures while gazing at the stars. And of course the inevitable speculations about extraterrestrial life.

Charbonneau cannot recall any such anecdote. But he does remember the moment that his scientific curiosity was aroused. He was about twelve years old and had just been to Expo '86, the World Fair in Vancouver in 1986. The Charbonneau family camped for a week in the Pacific Rim National Park on Vancouver Island. 'I spent the whole week playing in the rock pools under the cliffs,' he tells me. 'They stayed full of water at low tide, and were teeming with living organisms that I had never seen before. We lived in Ottawa, a long way from the ocean. The life in the sea continued to fascinate me, and I gave talks at school about sea urchins and anemones. If you had asked me in tenth grade what I wanted to be, I would have said a marine biologist.'

In his final year, however, Charbonneau became intrigued by physics and astronomy, after reading Stephen Hawking's *A Brief History of Time*. 'Hawking talked about the theory of relativity, black holes, quantum mechanics . . . I didn't understand much of it, but it did give me a glimpse of something magical below the surface. I thought, that's what I want to study. My mother encouraged me – she saw that I would probably be happier in physics than in biology.' When he went to study at the University of Toronto, Charbonneau mainly took subjects in astronomy. 'It was clear to me that there were still many fundamental questions unanswered in astronomy,' he says, 'while physics seemed much more advanced. I was afraid that, as a young researcher, I would only work in physics on the details of a problem that had largely been answered.'

Around that time, Sara Seager, an old study friend, had completed her doctorate at Harvard. She persuaded Charbonneau to apply for a place on the doctorate programme at the prestigious university. 'It had never occurred to me to go to Harvard,' Charbonneau says. 'I had grown up in Ottawa, and only had an image of Harvard from the movies. I knew Sara from the university hiking club. She is a very avid hiker and canoeist, much more than I am. She was a couple of years above me, and I looked up to her.' Charbonneau applied to Harvard and was accepted.

Doctoral students at North American universities usually start by taking a broad package of subjects and choose a specific research topic after a couple of years. When he arrived at Harvard in the mid-1990s, Charbonneau had already set his sights on a cosmology project. 'Cosmology, the study of the structure and history of the universe, was the most exciting discipline. Everyone at the university was talking about it,' he says. But he ended up taking a different direction, thanks to Bob Noyes, the professor who had been in the front row at Mayor's presentation in Florence. During a lecture, Noyes told his audience about the discovery of the first exoplanets. 'A handful of researchers from different disciplines – spectroscopy, stellar pulsations – had started working in this uncharted territory,' Charbonneau tells me. 'Plus, the question of whether extraterrestrial life existed was inspiring. I decided to take that step into the dark, too.' He decided to work on one of Noyes's projects, which focused on observing planets around other stars.

In the months following the first discoveries, there was still a fierce debate raging about whether exoplanets were real. Some critics firmly believed that the back-and-forth motion of certain stars was not caused by the gravity of a planet but by stellar pulsations. Others pointed out that the masses calculated for the planets were actually *minimal* masses. A measurement of the back-and-forth motion of a star using the Doppler shift only tells us about the star's movement along the line of sight, that is, towards us and moving away from us. It is, however, perfectly possible that the orbit of a planet is not exactly parallel to our line of sight to the star, and that we see it at a slight angle. Consequently, part of the star's motion is not observed via the Doppler method and it is actually moving faster than the speed we measure. A higher velocity means a more massive object; what we think is a planet may in fact be a much more massive brown dwarf.

The mass of an exoplanet can only be confirmed by observing a transit. Only then is it certain that its orbit is parallel to the line of vision and that its mass is small enough for it to be a planet and not a brown dwarf. Much more credible would be the discovery of a planet using both the Doppler and the transit methods. If exoplanets could be detected by two independent methods, their existence could no longer be called into doubt. Noyes believed that looking for transits was a very promising strategy. He told Charbonneau about a new telescope that a colleague of his, Tim Brown, was building for that very purpose in the Rocky Mountains.

TIM BROWN, the other man in the hut in the car park, is an experienced telescope builder. I spoke to him in 2013 by telephone, the most practical option given his nomadic existence. He works both in California and in Boulder, at the foot of the Rocky Mountains in Colorado. I heard a calm, deep voice on the other end of the line. There is a picture of Brown with one of his telescopes on his institute's website. It shows a middle-aged man with short brown hair and an impressive moustache.

He grew up in a university town in the north of Texas. 'I am a child of the Sputnik generation,' he tells me. 'I was seven years old when the Soviets sent their first satellite into space. There was a feeling in the U.S. that we were lagging behind in the space race. Science was in the air. I can never remember a time when I wanted to do anything else than astronomy.' On clear evenings, he would look through the telescope his older brother had built. Despite the glow from the lights of Dallas on the southern horizon, there was plenty to see. 'I was only vaguely interested in the planets in the solar system,' he says. 'I liked to look further away, at nebulae, at other galaxies. I remember thinking "Gosh, could there be someone looking at me from a planet somewhere out there? What would the Milky Way look like from there?"'

Before he was ten years old, Brown had read books by Arthur C. Clarke and Robert Heinlein about space travel and extraterrestrial civilizations. In Clarke's story *Earthlight*, there is an observatory on the Moon which, among other things, searches for exoplanets. 'I used to think sometimes, will it be possible to find planets around other stars in my lifetime? But I didn't spend a lot of time on it. I thought it was out of our reach, technically.'

He built his first telescope a few years later, getting most of the know-how he needed from books. His father, a professor of English, had close contact with colleagues in the technical disciplines and one of them donated a telescope stand that had been designed by his students. Brown ground the concave mirror himself. The most expensive part of the telescope was a single drop of aluminium, used to apply an ultra-thin reflecting layer to the mirror. Brown responds with humility when I express my admiration for his building project, saying 'Oh, building telescopes was a pretty normal hobby. Though I must admit I was the only one in my circle of friends who did it.'

After studying physics on the East Coast, he went to the University of Colorado in Boulder to do his Masters. Here, like many other students in the 1960s, he went crazy about theoretical physics. But a lecture on stellar atmospheres, which he had signed up for on a whim, brought him back to astronomy. The lecture was given by the renowned professor Dimitri Mihalas. 'He gave a brilliant lesson,' Brown tells me. 'I suddenly remembered what was so interesting about stars. This was physics, too, but the kind that appealed to me. It wasn't about abstract concepts that could only be understood in theories and formulas. It was about celestial objects, mass, pressure and radiation. Things that really exist. I was sold on it immediately.'

Astronomy became Brown's discipline, but he realized that his real passion lay in designing and building instruments. He chose an ambitious project for his doctoral research – making an instrument that would block sunlight, to test Einstein's theory of relativity. The theory stated that the Sun's gravity worked like a kind of lens, bending light. Seen from the Earth, stars behind the Sun seemed to shift if the Sun passed close to them. The experiment had first been conducted by the British astronomer Arthur Eddington in 1919, during an expedition to the island of Príncipe, on the west coast of Africa. During a solar eclipse, Eddington observed that the stars close to the Sun had indeed shifted position, thus confirming the theory of relativity and making Einstein world-famous overnight.

More than half a century later, scientists are still working on confirming the theory of relativity as accurately as possible.

That was the purpose of Brown's instrument, which was a kind of mega-accurate sun screen. The screen would shut out the glow from the sunlight visible around the silhouette of the Moon during a solar eclipse. That would make it easier to see the stars in the background and Eddington's experiment could be repeated with greater precision.

The screen was placed in front of the telescope and exactly covered the Sun. But then something strange happened: the Sun's diameter was not constant, but grew larger and smaller from hour to hour. The variations were minimal, around one ten-thousandth of the Sun's diameter. From that moment, all attention was focused on these fluctuations. They proved to be pulsations on the Sun's surface. This was not a new discovery; the phenomenon had first been observed in the 1960s, but there were no detailed measurements.

As with every star, two forces in the Sun are constantly battling with each other. Gravity draws all material to the centre. The reason that the Sun does not implode is that great pressure builds up in its interior that allows the nuclear fusion of hydrogen – forming helium from hydrogen atoms. This process releases enormous energy, generating an outward pressure that counteracts the gravitational pull. Gravity and the pressure of the nuclear fusion keep each other more or less in equilibrium and determine the size of the Sun. But, as the two forces are not exactly the same, they fluctuate around a middle point like a tumbler toy that has been given a push. And that is why the Sun is sometimes a little larger or smaller.

Brown abandoned the original purpose of the experiment, which was to measure the positions of the background stars. It became completely irrelevant anyway a few years later when space satellites proved able to measure the extent to which the light bent more accurately. For thirty years, the pulsating Sun was the only star that Brown studied and he became increasingly skilled at building instruments to measure solar pulsations.

'The first project was a revelation. It is crucial to work out in advance whether your instrument is accurate enough to answer your questions. In the modern age, you can't afford to build something and then conclude later "whoops, that isn't working,"' he tells me. 'No matter how tempting it is to get straight down to work, time, money and the competition force you to plan properly.'

In the first instance, Brown had measured the change in the Sun's diameter. But it proved much better to use the Doppler shift to record the pulsations in the centre of the Sun, rather than at its edges. During a pulsation, the surface of the Sun expands in all directions and the centre seems to move closer towards us and then further away. That makes the sunlight alternately bluer and redder, which can be seen in a shift in the absorption lines in the spectrum. Brown designed a special spectrograph to measure this shift.

It was no more than logical to assume that other stars would also pulsate, though that was more difficult to measure. The Sun is very bright, and its absorption lines are clearly visible. The fainter the light source, the less visible the lines are. In the early 1990s, together with his colleague Ron Gilliland, Brown focused on a number of bright stars, successfully using his spectrograph to measure their pulsations. Their method was in principle the same as that used by Mayor, Queloz, Marcy and Butler. Some of those who heard Mayor's presentation in Florence also interpreted the wobble of 51 Pegasi as a stellar pulsation.

After they had made a series of successful measurements, they ran out of stars that were bright enough. 'We ran into a brick wall,' Brown tells me. 'Our spectrograph could not measure fainter light sources.' The sensitivity needed for that was not to be achieved until a few years later, by the planet hunters. The strategy of recording the pulsations by taking light measurements – the brightness of the stars would after all also vary during pulsations – had also proved a dead end. The team had to find a new target, a new scientific question to tackle with their arsenal of telescopes. The question presented itself when Brown and Gilliland were invited to a workshop by a man with a mission: Bill Borucki.

'We had attracted Bill's attention with our plan B,' says Brown. 'Our work on light measurements. Kepler wasn't yet called Kepler then, but FRESIP. The proposal that Borucki's team had submitted to NASA had twice been rejected. Now they wanted to improve the light-sensitivity of their instruments.' At the workshop, Brown listened to the presentations and talked to the members of the FRESIP team. Their ambition to find planets outside the solar system inspired him: 'Ron Gilliland, Bob Noyes and I thought, you know what, we'll have a lot more fun if we look for exoplanets instead of stellar pulsations. In 1995, we switched over almost completely.'

The team had been working on a new exoplanet project for a couple of months when, during the night of 6 October, Brown got the call from his teammate Bob Noyes in Florence and answered with '*Buon giorno*, what's up?' Noyes told him what Mayor had shared with him in confidence about the forthcoming press conference at which the discovery of the planet around 51 Pegasi would be announced. Brown had already observed 51 Pegasi a few times himself and had noticed strange Doppler shifts. But he had interpreted them as measurement errors because, at the time, he had been adjusting the new spectrograph. 'Like everyone else, we were expecting orbital periods of years, not days,' he says.

Brown confirmed the existence of 51 Pegasi b with a series of extra measurements and sent a telegram around to the astronomical community. Then Brown and Noyes redoubled their efforts to find exoplanets. Instead of weekly, they now searched daily. And their research programme was given a new, catchy name: Stellar Astrophysics and Research on Exoplanets, STARE for short ('A bit of a contrived acronym,' Brown admits). In the first few years, the team found a grand total of one new exoplanet

– another hot Jupiter. Competitors like Marcy and Mayor were way ahead in the race and had better spectrographs.

He was getting no further with spectroscopy and the Doppler method, but Brown had an idea that might give him the edge over the other groups. It was inspired by his contact with the Kepler group, who wanted to observe planets passing in front of their stars. Brown wanted to observe the transit of a planet that had also been detected using the Doppler method. This combination of measurements would allow all the most important properties of the planet to be determined, including its mass, size and orbital radius (the distance between the planet and the star). The mass and size would give its density, revealing whether it was a gas or a rocky planet. The orbital radius, together with the known temperature of the star, enabled the temperature of the planet to be estimated.

'After the discovery of 51 Pegasi b, more and more Jupiter-like planets with small orbits were found,' Brown tells me. 'The unexpected existence of planets like this meant that we could take light measurements from the ground that were previously believed impossible. A Jupiter in transit causes a relatively strong stellar eclipse of 1 per cent. We could easily detect that with our equipment. What's more, we thought at first that Jupiters would have orbits taking ten years or more, much too long to for doctoral students, who have to come up with results more quickly, to work on them. But four days, that was a different matter.'

While Bill Borucki had to send his sensitive lightmeter into space to seek out the Earth-like planet he was so determined to find, Brown could in principle observe hot Jupiters from his back garden using a regular telescope. And that's just what he did. He built the telescope for the transit campaign himself, just as he had in his teenage years, including grinding and shaping the concave mirror. His friend Tom Baur let him use his chicken run as a workshop and he put the telescope together on the grass field next to it. The mirror was only 10 centimetres across – Isaac

Newton was already making larger ones back in the seventeenth century. The final model was set up in the wooden hut with a roof that could be hinged open, on the car park of Brown's research institute in Boulder. It was ready to search for transits, and an ideal project for an enthusiastic doctoral student.

AND SO it happened, in the summer of 1999, that David Charbonneau bought a second-hand car and loaded it up with his most important possessions for a road trip to the west, to the Rocky Mountains. 'I had no idea what would happen if the research project was a success,' he tells me. 'For me, it was above all an adventure. I had never been in the west of the U.S. before.' In Boulder, Charbonneau quickly made friends and found a house where he would ultimately live for what he calls 'a memorable year'.

During my own trip to the U.S., I repeat the last 100 kilometres of Charbonneau's journey in a hired Dodge. I, too, feel elated. The highway runs through an undulating landscape. Under a clear blue sky, I see the snow-capped peaks of the Rockies rise up in the distance. The local radio is playing John Denver's 'Rocky Mountain High': 'He was born in the summer of his 27th year, coming home to a place he'd never been before . . .'.

Boulder is a university town lying in the foothills of the Rocky Mountains. On the campus of the University of Colorado, there are buildings of sand-coloured stone, all with the same red roof tiles. The centre looks European and there are even hippy street musicians on the square in front of the old courthouse. A mountain river runs parallel to the main street. I park my car in the car park of Tim Brown's research institute and see the small, yellow-painted hut in which his telescope stood.

When Charbonneau arrived here fourteen years ago, he immediately had a plan for the first night of observations. Shortly before leaving Harvard, he had been given a golden tip by David Latham, who had found the first 'maybe planet' ten years earlier with Michel Mayor and the Israeli astronomer Tsevi Mazeh. In early 1999, with the same two colleagues, Latham had used the Doppler method to find what was very probably another planet, with an orbital period of three and a half days, which he believed was very likely to pass in front of its star. The article had not yet been published, but Latham told him the name of the star – HD 209458 – and predicted the moments at which the transits would occur. Tim Brown thought it a good idea to start with this star. 'We were still testing the telescope anyway, so why not test it on a star that might be interesting?' The two researchers tracked HD 209458 for ten nights with the telescope in the wooden hut.

The brightness of a star cannot be directly measured with a telescope. The telescope photograph first has to be processed by a computer, corrected for the known, undesirable effects of the instrument and compared to reference stars whose brightness has already been determined. This measuring method, known as photometry, is a standard job that an experienced observer like Tim Brown can pretty much do with his eyes closed. If he had processed the measurements from HD 209458 using photometry, which would have been about a day's work, he would have had a scoop.

But photometry is the kind of job that often gets left until later if there are more urgent things to do. In mid-September, Brown received a whole pile of measurement data from the Hubble Space Telescope from his colleague Ron Gilliland. The space satellite had observed a globular cluster, a spherical collection of an enormous numbers of stars in a relatively small area. For eight nights, it recorded the brightness of no fewer than 30,000 stars simultaneously, using the same strategy that Bill Borucki had in mind for the Kepler satellite. 'We gave this massive volume of data priority,' Brown tells me. 'Somewhere among those tens of thousands of stars, there must be one with a planet showing a visible transit.' The test observations of HD 209458 disappeared under the pile.

Ultimately, they did not find a single transit in the Hubble data. With hindsight, that was not so strange; stars in globular clusters are so closely packed together that their gravity drives each other's giant planets away. Large, easily detectable planets are therefore rare in globular clusters. After they had processed all the Hubble data, the pile of work had shrunk a little, but the measurements from HD 209458 would still have to wait for a few more weeks. Charbonneau had other things on his mind. In Boston, his empty desk had attracted attention and he was called in to explain himself. 'The people from the study committee at Harvard thought I had gone to Colorado to ski,' he says. 'When I told them about the planet project, they interrogated me for a long time. What would happen if we didn't find a transit? There would still be no useful research results for me to use for my PhD.' Charbonneau convinced the committee that, if that happened, there would still be enough material on stellar pulsations for him to write something interesting. He returned to Boulder.

Around 10 November, Brown and Charbonneau finally got around to looking at the brightness measurements for HD 209458. And it was immediately clear that they had seen two transits from the car park. During the transits, the star had become fainter by a few percentage points for two and a half hours. The first transit had occurred on 9 September and the other exactly a week later, on the 16th. That tied in perfectly with the period of three and a half days measured by Latham and Mayor's team (with the latter team having provided most of the measurements); the transit between the other two had taken place during the daytime. The depth of the transits suggested that the planet was 1.3 times larger than Jupiter. Charbonneau and Brown had proof that the planet, now named HD 209458 b, passed exactly in front of the star. That enabled its mass to be confirmed as a little less than half that of Jupiter. It was the first time that both the mass and the size of a planet outside the solar system had been determined, and meant that its density could also be calculated. HD 209458 b had to be a gas giant, like Jupiter and Saturn, but the gas it was made up of was much less dense. 'We slapped each other on the back and opened a bottle of Scotch,' Brown tells me. 'And then the telephone rang. It was Geoff Marcy.'

The fact that the telephone rang at exactly the same moment that Brown and Charbonneau fell elatedly into each other's arms may be a slight exaggeration. There are many different versions of the story: in one, Marcy left a message on the answering machine, in another he sent an email. But one thing was for sure, Marcy's team had observed a star with planetary transits. Brown told him in guarded terms that his team had discovered a similar star. As if they were bluffing each other at poker, the two astronomers finally laid their cards on the table: they had both discovered the planet around HD 209458.

The star had been on Marcy and Butler's observation list for a few months. A few days before their call to Brown, Butler had discovered a wobble in the star's spectra over the preceding months. He assumed that a planet would also display transits and, on the basis of their measurements, he and Marcy were able to work out exactly when they would occur. Marcy immediately contacted Gregory Henry, a light measurement expert who worked with a telescope in Arizona. At the moment Marcy predicted the transit, just after sunset on 7 November, he pointed his telescope at HD 209458. When he processed the data the following morning, Henry indeed saw that the star had become a few per cent fainter in the final hour. After that, however, it had disappeared below the horizon, so that the end of the transit had not been seen. The next transit, three and a half days later, would occur during the day and not be visible with the telescope.

Marcy was unsure what to do. Half a transit was not enough to publish their findings. On the other hand, he didn't want to miss out on the scoop, as he had with 51 Pegasi. He settled for a compromise: a telegram to the astronomical community announcing that the star had been slightly fainter for an hour. He called for more observations to be made at the times he had predicted, to confirm the transits. The following day, before the telegram had appeared on the Web, he had personally informed a few colleagues, including Tim Brown.

Brown and Charbonneau jumped out of their skins. Unlike Marcy and Henry, they had two complete transits. They had hardly hung up the telephone when they started writing an article to get the news out before Henry, Marcy and Butler. A few days later, they were shocked again when Berkeley issued a press release on HD 209458, in which the discovery was attributed solely to Marcy's team. The final, bittersweet detail in this drama is that both articles were sent to the same publication, the *Astronomical Journal*, on the same day. Both were published in the same issue, but the one by Henry, Marcy and co. was first. In the history books, both teams are given equal credit for detecting the first transit.

The million-dollar question is, of course, how coincidental was it that the star that Latham had told Charbonneau about a few months earlier had ended up on Marcy's list? There are thousands of stars just as bright as HD 209458. How did this one find its way onto two observation lists within a couple of months? Who had tipped off David Latham and Geoffrey Marcy? All those involved are vague about it, or say that they really don't know. None of them seem to find the issue important. 'It is good that it ended up with the honours shared,' says Charbonneau. 'At the time we weren't happy about that press release, but Marcy has apologized for that. We have no hard feelings about it. After all, they did discover it independently.'

IT IS raining in Boulder. I drive to the hut where Brown and Charbonneau made their observations in 1999. I don't need to look for long. Behind the car park is a small lake; there are few houses, so there is little artificial lighting at night. In front of the lake, there are two tiny, light-yellow huts. The one on the left is a little higher, but you still need to watch your head as you go in. There is a simple padlock on the door. A couple of pallets and orange bollards are stacked up next to the hut.

I look around me. The air is cloudy. To the west, there are almost no trees to be seen. On a clear night, and if the lights of the institute are out, the conditions for observing the stars are good, but not great. The telescope hut is inconspicuous, half the size of an average garden shed. It is totally incomparable to the gigantic telescopes I have seen in Chile, in the Mars-like landscape of the Atacama Desert, where it is dry and crystal clear every night. And yet it is still possible to defeat these Goliaths with simple equipment – from a hut in the corner of a car park and with a telescope you put together yourself. Both Charbonneau and Brown still make frequent use of small telescopes for their research. They lead their own worldwide network of telescopes, some of which are operated by amateur astronomers, who are working together to search for exoplanets close to the Earth.

IN LEIDEN, Charbonneau looks back on his year in Boulder. 'I had no concept at all of the consequences, because I was still a rookie,' he tells me. 'Geoff, Tim and the others knew how important our observation of the transit of HD 209458 b would be. It was a crash course in doing exciting science. I was very lucky to be at the focal point of that.'

It was indeed a milestone. An exoplanet had been observed using two independent methods. There was no more denying it: there were planets outside our own solar system. Speaking of the transit, Marcy told the BBC that 'for the first time in human history, we have confirmation of a planet orbiting another star.' And observing the transit was simpler and less abstract than the Doppler method. Didier Queloz told me that his colleagues in Geneva now finally dared to look him and Mayor in the eye again, after doubting the existence of exoplanets – and therefore the scientific integrity of their discoverers – for so many years.

Bill Borucki, too, was over the moon. He pointed his Vulcan Telescope at HD 209458, observed a transit and persuaded NASA that the instrument he wanted to fit on board the Kepler satellite would work. CoRoT, a European equivalent of Kepler, was also given the green light to search for more transits. The telescope was launched before Kepler and – ironically enough for Borucki – discovered the first exoplanet from space in 2007.

THE QUESTION of whether exoplanets existed had been answered definitively by the discovery of the two American research teams. Three years after detecting the transit, Charbonneau had another scoop. He was helped by Sara Seager, the university friend from Toronto who had talked him into going to Harvard. She had developed a method of using planetary transits to find out more about something even more interesting, the planets' atmospheres.

Seager got the idea from her doctoral supervisor at Harvard, Professor Dimitar Sasselov. Originally from Bulgaria, Sasselov had settled in North America after the fall of the Soviet Union. He had attended the conference in Florence in 1995 and was astounded to hear about Mayor's discovery of the first exoplanet, using the Doppler method. Many years later, Sasselov still remembers it clearly: 'I was incredulous – the period was so short, it was measured in days, not years – I told my wife back in the hotel that night – just 400 days!' The following day, he proved to have been mistaken: the period was only 4.2 days. 'The night before, I must have heard "4.2 days", but being so incredibly foreign to my preconception, my brain had "translated" that number to a more "reasonable" 420 days, or – roughly 400. Deeply held preconceptions can be very powerful.'

Sasselov became more and more fascinated by the planets that were being discovered here and there. He had gone to Florence because of his expertise in stellar atmospheres – the outer shells of the hot globes of gas. Hot Jupiters like 51 Pegasi b were most probably also gas globes with a rarefied outer atmosphere. If such a planet passed in front of its star, the starlight would shine through its atmosphere on the outer edge of its silhouette. The greatest part of the light would pass through the atmosphere unhindered but, at certain wavelengths, it would be absorbed. Those wavelengths could be recognized as the 'fingerprint' of the different atoms in the atmosphere. The spectrum of the planet's atmosphere would therefore contain absorption lines, just like the spectrum of the star itself.

In 2000, Seager and Sasselov published an article in which they predicted that the element sodium would cause one of the clearest absorption lines. They proposed a method for detecting this line in the case of HD 209458 b, the planet whose transits Charbonneau and Brown had measured shortly earlier. The spectrum would have to be observed twice, once without the planet and once during the transit, as it passes in front of the star like a kind of photobomb. The first time, only the starlight is received, and the second time it will also contain the planet's fingerprint. If you deduct the spectrum of the first from the second, only the spectrum of the planetary atmosphere will be left.

Charbonneau and Brown, the observers, were happy to take up the suggestion of their colleagues the theoreticians. In 2002, they successfully applied Seager and Sasselov's method using the Hubble Space Telescope. They saw that a very small part of the starlight was absorbed during the transit, exactly on the wavelength of the sodium atom. This was the first time the composition of the atmosphere of an exoplanet had been observed.

Three years later, Charbonneau was again one of the first to successfully apply a new observation method. But this time, he not only measured an absorption line, but picked up the light from the planet itself. Every object, or 'body', radiates light; not only celestial bodies, but human ones. And the colder the body, the longer the wavelength of the radiation. A star is hot and radiates visible light. Our own bodies are much colder than a star and therefore radiate light at a much longer wavelength – infrared light. That is why human bodies can be seen in the dark with infrared cameras. Hot Jupiters are warmer than people, but cooler than stars: they, too, radiate infrared light.

By coincidence, a satellite was launched in 1993 that could detect this radiation from space – the Spitzer Space Telescope. Spitzer was not designed to observe exoplanets – they had not yet been discovered when it was being developed – but the planet hunters were happy to make good use of it. Two competing teams, one led by Charbonneau and the other by Sara Seager and planet hunter Drake Deming, focused the space telescope on two different stars. They took two infrared photographs of each star, one with the planet in front of it and one with it behind. They laid the two images on top of each other and removed all the corresponding data, deducting – as it were – the one from the other. What remained was the absorption spectrum of the exoplanet atmosphere. Both teams published their results in the same year.

CHARBONNEAU IS a man who likes to be at the forefront. Besides notching up the first transit, the first exoplanetary atmosphere and the first detection of the infrared radiation of a planet, he also had a hand in more recent discoveries. He responds shyly to the question of whether he has a nose for a scoop. He doesn't deny it, but explains why his discipline is so suited to such groundbreaking discoveries: 'In this new field of research, even the basic questions haven't been answered yet. How often do planets occur? Is the structure of the solar system unique? Does a second Earth exist? There is no one who has done forty years of research in this field. Not a single exoplanet expert has a position of authority based purely on experience. Yes, Bill Borucki has been in the business for a long time, but even he can't tell us what the real opportunities are to find a second Earth – our knowledge is changing too quickly for that. It is only when you make new discoveries that you see new opportunities.

'In such a new field, even a young PhD student like myself could make an important discovery. And that's how it should stay. Policy-makers always want roadmaps, advance planning to determine what missions will get funding in the future. There is a great temptation to try and predict the future and say "we will need this instrument in twenty years", and then devote all resources to working on it. In those twenty years, you can better invest in research scholarships for young people. Young researchers with a fresh way of looking at things, who can innovate and do "aggressive" things. They are the ones who should lead the big projects. If we don't give them that chance, they will not be encouraged. They will keep their mouths shut, and we won't benefit from their talent. I hope I can offer my students the same kind of opportunities as I got in Colorado.

'I try and look at the resources available to do something for the first time, to come up with a new observation method, that my students and I can demonstrate for the first time. Once you do that, a lot of smart people jump on it with money and other resources. And then we move on to something else.'

For Charbonneau, Brown, Seager, Marcy and all the other planet hunters, there is one clear end goal on the agenda: to find an answer to that one question that is even older and more fundamental than the one about planets outside our solar system. The question that was asked before people even knew what planets and stars were. The question of whether there is life out in space. The answer starts with a first step: finding a habitable planet.

eight

GOLDILOCKS AND
THE RED DWARFS

THE LIFEGUARD sees her immediately through his binoculars. A young woman splashing around in the water while, 20 metres away, a shark's fin moves towards her. Without hesitating for a moment, the man sprints into the surf, gives the shark a few hefty punches and lifts the dazed maiden out of the water. She comes round in his arms, gazes longingly into his eyes, her mouth moves towards his, and then . . . her eyes shift to something that is moving off screen. Her mouth falls open. The camera follows her stare. A cumbersome figure in an enormous white suit and a reflecting helmet comes walking towards them, his boots sinking into the loose sand. It is an astronaut. The bikini babe works herself loose from the grip of her saviour and runs towards the white apparition. The words appear on the screen: *Nothing beats an astronaut.*

The slogan in the ad for Apollo, the new deodorant from the Axe/Lynx line, is true. For many years, eight-year-olds have put 'astronaut' at the top of their list of best jobs, together with fireman, animal sitter and superhero. When she was that age, Elisa Quintana, researcher at NASA and the SETI institute, had no interest at all in astronauts or astronomy. Most of all, she wanted to become a ballerina or a drummer ('Actually, I still do'). After graduating from high school, she went to junior college, where – not knowing what to do – she chose a broad package of subjects. She discovered that she was good at mathematics and physics. But it was not until she was studying physics at the University of San Diego that a meeting with a real astronaut changed her life for good.

Sally Ride was not just any astronaut; she was the first American woman in space. She was part of the otherwise all-male crew of the space shuttle on two missions. Modest by nature, Ride reluctantly became an important role model for young women with an interest in the natural sciences. Space was the perfect environment to show once and for all that the widely held prejudice that science is a man's business was wrong. 'Weightlessness is a great equalizer' is one of her inspiring statements. After her two missions, she became a physics professor at San Diego. There, she launched KidSat, which gave children the opportunity to take photographs of the Earth using a camera on the space shuttle. Elisa Quintana was one of the students who helped Ride on the project. Quintana worked out at what moment on the flight route the requested photos had to be taken.

'That's how I was, as it were, launched into space,' a cheerful Quintana tells me in a video conversation. She didn't become an astronaut, but her work with Sally Ride put the seal on her choice of study. 'Since then, I've actually only studied planets and stars,' she tells me. 'Especially how multiple planets orbit a single star and how their orbits can change over a long period.' But Quintana didn't want to spend all her time on theory; she wanted to take part in a real space mission and make observations. She got that chance while she was researching her doctorate. She met Bill Borucki, who was at that time fighting for his Kepler satellite, and, in 1999, became one of the observers using the Vulcan Telescope in California to test the CCD camera that was to be placed on the satellite. She has now been a member of the Kepler team for fifteen years and works at the NASA Ames Research Center. 'I recently moved to an office with windows,' she says. 'Now I look out on the runway, where all kinds of aircraft take off and land all day. I even saw Air Force One here once.'

The reason I am talking to Quintana in May 2014 is her discovery a few weeks earlier, which made headlines around the world. The *New York Times* ran an article headed 'Scientists Find an "Earth Twin", or Perhaps a Cousin'. Other reports spoke of a 'potentially habitable planet', the 'Holy Grail' of planet hunters. Quintana laughs bashfully when I ask her if her find will go down in the history books as the discovery of the century. Unlike Geoffrey Marcy, who thrives on being in the spotlight, she is somewhat shy and modest. She even seems a little overwhelmed by all the attention her research has attracted. 'The planet we found is special, but is by no means the end of our search. I see it more as one of a series of milestones we have achieved in recent years.'

The milestones that Quintana is referring to are the discoveries of exoplanets. Each time a new planet is discovered, there is considerable attention for any properties that resemble those on Earth. Every 'first' (the smallest exoplanet to date, the first with water and so on) is another step towards finding the answer to the ultimate question: whether there is life on a planet other than the Earth. To answer that question, we have to know how we will be able to recognize that life. We only have one example of life in the universe, that on Earth. We use that as a standard to compare with other planets. The first questions that need to be asked, therefore, are what conditions make life on Earth possible, and can we identify these conditions on exoplanets that are many light years away from us?

IN ANCIENT cultures, people had already observed that most things on Earth were made up of, or were dependent on, four elements: fire, earth, water and air. References to these elements crop up everywhere, from Babylonian clay tablets to Greek myths, from Egyptian hieroglyphics to Japanese haikus, from horoscopes to disco music (Earth, Wind and Fire). Modern writers, too, make use of the symbolism of the four elements. In Dan Brown's book *Angels and Demons*, the murders of four cardinals are associated with them (one is burned alive, another chokes on earth and so on). Even the houses in Hogwarts School of Witchcraft and Wizardry in J. K. Rowling's *Harry Potter* books are derived from the four elements. And then there is the novel series *A Song of Ice and Fire*, which gave rise to the phenomenally successful television series *Game of Thrones*.

In scientific terms, the four classical elements have long been obsolete. There are now more than a hundred elements in the periodic table, which are themselves made up of even smaller component parts – electrons, protons and neutrons – the last two of which are comprised in turn of minuscule quarks and gluons. Yet there is good reason why our forefathers attached such importance to the four elements, as they symbolize the most important things that make life on Earth possible. Fire is a source of energy and earth is the solid ground beneath our feet, while we need water to drink and air to breathe to stay alive.

In science, the element of fire is represented by the stars. In the interior of the stars, nuclear fusion takes place, a process that produces energy at more or less the same level for billions of years. The light from the stars shines onto planets and warms them up. On Earth, plants use that light for photosynthesis, which produces oxygen. Another reason why stars are so important is their great mass. Because of the gravitational force this generates, planets continue to orbit their stars seemingly endlessly. That creates regular seasons on the planets and a relatively constant temperature, which is favourable to the development of life.

The second element that makes life possible is earth. Living organisms have to be able to move over land or through water on a planet with a solid surface. That means we are looking for rocky planets, like the Earth and Mars, rather than gas planets like Jupiter and Saturn. Rocky planets are small: a planet as big as Jupiter cannot consist entirely of rock. It would be so massive that it would partly collapse under its own gravity.

Water is the third element that is crucial for life on Earth. It is a good solvent for the carbon molecules that make up all living organisms on the planet. It thus works as an efficient means of transporting carbon compounds. Blood, for example, which is mostly made up of water, transports nutrients and waste products through the human body. We drink water, we consist largely of water. Practically every organism on Earth depends on water.

An important extra condition is that the water must be in liquid form. At temperatures above 100°c water evaporates, and below zero it freezes. Whether liquid water can exist on a planet depends very much on how far it is from its star: the closer it is, the hotter the planet. Mercury and Venus are nearer to the Sun and are therefore hotter than the Earth. On these planets, water evaporates. On Mars and the planets further away from the Sun, water only occurs as ice. Around every star, there is a 'habitable zone', where the presence of liquid water is, in theory, possible.[23] This area is also known as the 'Goldilocks zone', after the famous fairy tale in which Goldilocks sneaks into the bear family's house and finds three bowls of porridge on the table. The first is too hot, the second is too cold and the third is just right. Planet hunters search for habitable worlds in the Goldilocks zone. In our own solar system, only the Earth and Mars fall within this zone (which is why we keep sending robots to the red planet to learn about conditions there). Strictly speaking, depending on which astronomer you ask, Venus also formally lies in the Goldilocks zone. But the planet's thick atmosphere creates an extreme greenhouse effect, causing temperatures on the surface to soar to an average of 460°c. This makes Venus too hot to live on, and shows how misleading the term 'habitable zone' is.

The fourth and final element is air, in this context the planetary atmosphere. The Earth's atmosphere contains many gases that are used and produced by living beings. Some of these gases, like oxygen and ozone, would not be in the atmosphere at all if there were no life. They are known as 'biomarkers', unmistakable indicators of life on a planet. Planetary atmospheres offer the best – and perhaps the only – chance of finding life on other planets. More of this later.

THE FIRST milestone in the search for the four elements else-where in the universe was the discovery of the first exoplanet, 51 Pegasi b, by Mayor and Queloz in 1995, followed shortly afterwards by a handful of others. By the turn of the century, some twenty exoplanets had been discovered. Most of them were hot Jupiters, gas giants in short orbits around their stars. The very existence of these 'barbecue planets' was astounding enough in itself, but a gas planet with a temperature of 1,500°C is not a pleasant environment for organic life. It did not, however, stop people from asking the million-dollar question: newspaper reports invariably ended with the claim that this was the 'first step' towards finding extraterrestrial life.

The next milestone was the discovery of exoplanets that were significantly smaller than the hot gas giants, the 'super-Earths'. The first planet of this kind, GJ 876 d, was found in 2006. It is the third planet around the star GJ 876 (like the prefix HD, GJ refers to the star catalogue that lists the star). These planets are approximately ten times bigger than the Earth, but are smaller than Neptune. In theory, they are small enough to be rocky. The European satellite CoRoT – which pipped the Kepler mission to the post – discovered another super-Earth in 2009, COROT-7 b (so called because it orbits the seventh star around which the satellite had discovered a planet). In that same year, David Charbonneau found a similar example, GJ 1214 b. This exoplanet was discovered using both the Doppler and the transit methods, enabling its size and its mass to be calculated. The planet's density resembled that of water, leading some astronomers to speak of an 'ocean planet', a gigantic 'water world'.

It was Dimitar Sasselov, the Bulgarian Harvard professor, who devised the name 'super-Earths' for this category of exoplanets. Some of his colleagues, however, were not happy with this term. Geoffrey Marcy still considers it misleading, as it suggests that the planets are similar to the Earth in terms of gravity, atmosphere and habitability. And that is not the case: many of the super-Earths have a thick outer layer of gas and water. In that sense, they more closely resemble Uranus and Neptune; for that reason, Marcy suggests calling them 'sub-Neptunes'. Sasselov defends his preference for super-Earth in an interview, saying 'We call stars that are bigger than giants, super-giants; we call stellar explosions which are more energetic than novae, super-novae; so it just made sense that if you have a planet which is larger than the Earth but otherwise is in essence similar to the Earth, you would call it super-Earth. I guess I didn't grow up with Superman.' If there is one thing that astronomers can never agree on, it is what to call their discoveries.

THE DISCOVERY of the first super-Earth in the Goldilocks zone of a star was another step forward. In 2007, the exoplanet team from Geneva announced the discovery of three planets around the star GJ 581. The second and third, designated as GJ 581 c and GJ 581 d, were super-Earths. Planet c was in the Goldilocks zone, at the right distance from its star to accommodate liquid water. One member of the Swiss team was bullish in a press release, saying 'On the treasure map of the universe, one would be tempted to mark this planet with an X.' In no time, T-shirts appeared with the text 'I'm off to see planet 581 c' and postcards showing a fantasy landscape and the message 'Greetings from GJ 581 c.'

A little later, a theoretical study was published that severely questioned the habitability of the planet. The distance from its star is not the only factor influencing the temperature on a planet; the composition of the atmosphere is also very important. Greenhouse gases like carbon dioxide (CO_2) act as a kind of blanket, warming the planet up (as in the case of Venus), while other gases can cause the starlight to be deflected, thereby keeping the planet cool.

The Goldilocks zone alone therefore does not offer a perfect guarantee of liquid water. The German authors of the study calculated that a dense planetary atmosphere around GJ 581 c would cause an extreme greenhouse effect, making it much too hot for life to survive. GJ 581 d, which was a little further away from the star, could have just the right temperature. However, the authors assumed that the planets had the same composition as the Earth, which was by no means certain, as only their mass and their orbits were known. This limited data was not enough to reach any conclusions regarding habitability. The properties of the planet itself – its density, composition, atmosphere, the topography of its surface – are at least as important.

THE GJ 581 system had more surprises in store. A year after the discovery by the Swiss group, a fifth and sixth planet were found orbiting the star. The last, GJ 581 g, was found by the American team of Steven Vogt and Paul Butler, which had since split off from Geoffrey Marcy's group. The discovery was spectacular: GJ 581 g was only three times more massive than the Earth and was in the middle of the star's Goldilocks zone. It was a world with just the right temperature for liquid water, where life like that on Earth could develop. This was the discovery that every planet hunter dreamed of. In the word of thanks, where authors normally mention only their colleagues and sponsors, Vogt also thanked his wife Zarmina 'for her patience, encouragement and wise counsel'. The official name of the planet may have been GJ 581 g, but for Vogt his discovery would always be known as 'Zarmina's World'.

Vogt and Butler announced their discovery at a press conference in September 2010. Terms like the 'Holy Grail' and 'the Earth's twin sister' were bandied about. Everyone was convinced that this was the climax of a search that had lasted hundreds of years and had gathered momentum in the past fifteen years. Of course, most of the questions were about the chances of there actually being life on GJ 581 g. Steven Vogt, normally a calm and rational scientist, allowed himself to make a very daring claim. 'Personally,' Vogt said, 'given the ubiquity and propensity of life to flourish wherever it can, I would say, my own personal feeling is that the chances of life on this planet are 100 per cent. I have almost no doubt about it.'

Astronomers are only people, too. You do your best to remain objective in your work, but sometimes it is impossible to switch off your emotions. There were dozens of curious and crafty journalists at the press conference, which was broadcast live on the Internet, who wanted a snappy quote for their articles. Perhaps this lead Vogt to speak from his heart rather than from his head. When Butler was asked if he was just as optimistic as his colleague, his response was a little more cautious. 'I hate to even speculate on how optimistic I feel about the chances of there being life on this planet,' he said. Later in the press conference, Vogt stressed that this was his own personal belief, saying that, though he was not a biologist, 'I've cleaned out my shower enough to know that, if there's any residual water left anywhere, life tends to find it pretty quickly.'

Vogt's insistence that it was his own personal feeling was of course lost when his words were repeated ad infinitum on blogs and by the media. 'Astronomer 100% Certain of Life on Other Planet' ran the headlines. American news stations spoke to mediagenic 'experts', who called on NASA to send a mission to GJ 581 g without delay 'just to see what is there!' Vogt's friends and enemies feared for Van de Kamp-like scenes: a scientist with a good reputation making statements that he could not back up with facts (except with home truths about his bathroom). This kind of publicity puts the whole of science at risk. If astronomers speculate willy-nilly about complex questions like the existence of extraterrestrial life without basing their claims on hard evidence, why should anyone take them seriously? One contributor to a scientific forum noted drily 'my own personal feeling is that the chances of Vogt regretting his statement are 100%.'

Rather than fanning new speculation about extraterrestrial life, other planet hunters wanted first to know whether Vogt and Butler's planet genuinely existed. Competing teams from Europe and the u.s. pointed their telescopes at star GJ 581. None of them were able to detect in their measurement data the subtle wobble that would betray the presence of the small planet GJ 581 g. The Doppler shift in the spectrum, which Vogt suggested was a planet, could not be distinguished from the interference caused by inaccuracies in the measurement apparatus. Vogt himself also analysed the data gathered by his opponents, but saw no reason to retract his discovery.

The final blow came in the summer of 2014, when a study published in *Science* definitively made mincemeat of both GJ 581 d and GJ 581 g. The light variations that had been attributed to these planets were in reality changes in the luminosity of the star itself, a little like the pulsations that Tim Brown had measured in the 1990s. Zarmina's World proved ultimately to exist only in the word of thanks of a lovestruck astronomer.

DESPITE THE milestones and the accompanying media attention, exoplanets remained a curiosity in the first decade of this century. Every now and then, one would be found orbiting a star. Some were small, others large; yet others potentially met the criteria for being habitable, but no one had any idea whether such planets were common or rare. Frank Drake had drawn up his equation in 1961, estimating the number of intelligent, communicating civilizations in the Milky Way. One of the unknown quantities in this equation was the average number of habitable planets around a star. That number could only be estimated after systematic research, when there were sufficient planets to correct for bias effects.

A bias is a systematic error, a distortion of the truth caused by the measurement method. One cause of bias can be searching only for planets where it is easiest to find them. Writer David Freedman called this the 'streetlight effect', after the following story: one night, a policeman sees a drunk crawling on his hands and knees under a streetlight. The man tells him he is looking for his keys. The policeman decides to help him and gets down on his knees, too. After searching for a few minutes, he asks the drunk if he is sure he lost his keys here, to which the man replies, 'No, I lost them in the park. But the light is better here.'

At first, planet hunters searched mainly under the street-light: they looked for planets that were the easiest to find. Of the approximately three hundred exoplanets that had been found in 2009, the majority were more massive than Jupiter. The Doppler method was especially good at detecting these planets: a massive planet causes a strong and easily measured wobble in its star. But that does not mean that most exoplanets are more massive than Jupiter. The general statistics for exoplanets – for example, how often a planet of a certain mass, size and temperature occurs – were still unknown.

That question was answered in the same year, 2009, when Bill Borucki's long-fought-for dream – the launch of the Kepler satellite – was finally fulfilled. As it takes longer to process data than it does to gather it, Borucki's great moment did not actually come until two years after the start of the mission. In February 2011, at a NASA press conference, he announced the results of the first four months that the telescope had spent gazing at the same area of the sky. The result was overwhelming: it had found 1,235 objects that were almost certainly planets, 54 of which were in the Goldilocks zone. The *New York Times* was jubilant: 'Astronomers have cracked the Milky Way like a piñata, and planets are now pouring out so fast that they do not know what to do with them all.'

When a second batch of 715 planets was announced three years later, in 2014, it was the NASA researchers themselves who provided the sensational texts, claiming that Kepler had found the 'mother lode' of exoplanets. By the summer of 2016, two more years of planet hunting had produced nearly three thousand confirmed exoplanets. In many cases, there were systems with multiple planets orbiting the same star. More than 2,500 'candidate' planets are still waiting in the wings for their existence to be confirmed by other observers.[24]

New pearls were quickly found among this enormous catch. This was the 'series of milestones' that Elisa Quintana was talking about. Every time there was an interesting find, the press were called together: Kepler-22 b, the first exoplanet the satellite found in the Goldilocks zone; Kepler-10 b, an Earth-like exoplanet (but too hot for life); Kepler-78 b, a planet with almost the same density as the Earth.

In April 2013, Kepler-62 e and f were presented with much pomp and circumstance. Both planets were only one-and-a-half times larger than the Earth and were located in the Goldilocks zone. The American House of Representatives summoned a number of NASA and SETI scientists to a special hearing and asked them 'Have we found other Earths?' That was, of course, not an easy question to answer. It is possible that such small planets are rocky, but they may also consist largely of water. To establish that beyond doubt, you also need to know their mass and the composition of their atmosphere. As usual, the scientists told the politicians: that requires more observation, time and money.

THE MAIN question that the Kepler Space Telescope was intended to answer – how many Earth-like, habitable planets a star has on average – has not yet been answered unequivocally. More super-Earths than Jupiters have been found, but we still do not know whether there are more even smaller planets. Kepler, too, is prone to bias effects. Earth-like planets are more difficult to find because they absorb only a small quantity of light during a transit. It is therefore very possible that Kepler actually does not see most Earth-like planets. The handful that have been found so far may just be the tip of the iceberg. On the basis of the planets that the space telescope has found to date, it is possible to estimate the size of that iceberg. How many planets has the satellite not been able to see? It was rock star and planet hunter Jon Swift who came up with the answer to this question.

Swift is a researcher at the California Institute of Technology, Caltech for short, the old university of famous names like physicist Richard Feynman and biochemist Linus Pauling, and a favourite hangout of Albert Einstein. I speak to Swift one morning on Skype. He lounges in his desk chair, relaxed. His appearance doesn't match the stereotypical sandals-and-socks, pen-in-breast-pocket and glasses-with-bent-frame image of a scientist. He has a neatly trimmed beard, wears a green army cap and a shirt with the top buttons undone. He looks like a surfer who sits on the beach with his guitar, singing songs about the ocean. And that's just what he is – a surfer/songwriter. He has brought out four albums and written songs for a number of 'iconic' surfing films. He's a kind of Jack Johnson with a PhD from Berkeley.

After completing his PhD, Swift devoted himself entirely to his music and toured the country with his band. The high point of the tour was a concert in the legendary Troubadour club in Hollywood. He had invited John Johnson – not a relation of Jack's but an old room-mate from Berkeley. Johnson was setting up a group of his own at Caltech to conduct research into exoplanets. When he came backstage after the concert, he managed to persuade Swift to come and work for him. 'I don't know why he offered me a job,' Swift tells me. 'Maybe he really liked the concert or the atmosphere backstage.'

Swift has two homepages. One is for his music, showing his albums and tour dates. It is artistically designed with photographs of mountain landscapes, pick-up trucks and guitars. Hidden among the links is one entitled 'Astronomy: Jon's alter ego', which takes the visitor to Swift's research page. On this smartly designed, grey page, you can download his academic cv and there are links to his publications. There is no mention here of his music. 'It's not that I want to keep anything secret,' Swift explains. 'The music world and science are simply two different parts of my life.' The only time that Swift agreed to play at an astronomy conference proved a disappointment for everyone involved. 'It just didn't work,' he says. 'The audience weren't there to listen to music, but for the science. They didn't really experience the music, it didn't get off the ground. You can feel that from the stage, you know.'

Swift's folk music may be appreciated by a select audience, but his work in astronomy focuses on the mainstream: red dwarfs, the most common kinds of stars. Drake's question – how many planets does the average star have? – actually starts with another question: what is an 'average star'? In the Milky Way, the rule of thumb is the larger the stars, the fewer there are. Three-quarters are red dwarfs: not half as massive as the Sun and thousands of degrees cooler. They also radiate much less light; even in the brightest night sky, you can't make out a single red dwarf with the naked eye. So when you look up at the sky at night, three-quarters of the Milky Way is invisible. Most of the red dwarfs in its field of vision were even too faint for the Kepler telescope. Only a few thousand stars of the 150,000 on Kepler's list were red dwarfs.

'We wanted to study the red dwarfs in the Kepler zone more closely,' Swift tells me. 'We had found a handful of planets around them, but we knew that Kepler had missed a large number of them. The satellite only detects planets that, from our perspective, move in front of their planets. And red dwarfs are less bright, which makes it more difficult to distinguish transits.' One red dwarf, Kepler-32, proved to be a goldmine. No fewer than five planets orbited the star, causing transits that alternated with each other in a complicated but regular rhythm that resembled a jazz drum solo.

Swift, Johnson and their colleagues used this system as a kind of Rosetta stone, a blueprint for the rest of the red dwarfs that Kepler had observed. They compared the orbits and sizes of the Kepler-32 planets with those of planets that had been found around the other red dwarfs. Then they made a rough but statistically sound 'back-of-the-envelope' calculation of the planets that Kepler had missed. Swift came to a conclusion that was adopted as a one-liner by the entire astronomical community: every red dwarf has at least one planet. A year after the article was published, Swift still stands by his claim. 'Other teams have repeated the analysis independently and come to the same conclusion,' he says. 'Planets, and especially small planets, are everywhere in the Milky Way.'

Journalists looking for good headlines immediately started doing their sums. Because red dwarfs are the most common kind of stars, Swift's conclusion was quickly interpreted as 'every *star* has at least one planet.' There are approximately 300 billion stars in the Milky Way; three-quarters of them are red dwarfs, and according to Swift's calculation, each has at least one planet. News channels like NBC and newspapers like *The Telegraph* rounded the figure down, declaring 'More than 100 Billion Alien Planets in the Milky Way!'

MOST ASTRONOMERS see these statistics as the most important outcome of the Kepler mission. 'Kepler brought about a revolution in what we know about planets,' says David Charbonneau. 'When Bill Borucki first started, we knew nothing. It could also have been possible that only one in a million stars have planets. There were no limits; everything was possible. Now we know that planets are not all rare. There may even be more planets in the Milky Way than stars. Kepler's legacy is the knowledge that planets, especially small planets, occur everywhere.'

A wonderful illustration of this legacy is the Kepler Orrery. An orrery is a working model of the solar system, like the one with the copper ball and the marbles that my grandfather had shown me in the Teylers Museum. The planets and the distances between them are made to scale and, when you turn a handle, they move around the Sun. These mechanical models were very popular in the eighteenth and nineteenth centuries, when new planets and moons were regularly found in the solar system, but they are now largely museum pieces.

That was to change when Daniel Fabrycky, a researcher on the Kepler mission, reintroduced the orrery in digital form after the great planetary haul of 2011. His orrery showed all the planetary systems that Kepler had found and which comprised more than one planet. After each new batch of planetary finds, Fabrycky updated the orrery online. Each planet was represented by a coloured dot on the screen, following its orbit along a white circle. Planets orbiting the same star are shown with different colours. Some systems have seven planets circling the same star. The film is a favourite at public lectures by planet hunters. The speaker can take a sip of water while his audience takes in what they are seeing. Hundreds of solar systems, with a wide variety of planets circling them in all permutations. Large planets in small orbits, small planets in large orbits. You name it, Kepler has found it.[25]

One of those spinning dots is Kepler-186 f, Elisa Quintana's discovery and the most recent milestone in the planetary harvest of Borucki's space telescope. The planet is only 1.1 times larger than the Earth and orbits its star within the Goldilocks zone. It is the smallest Goldilocks planet found so far and the first of its kind around a red dwarf. 'That's why our discovery has caused such a furore,' says Quintana. 'Three-quarters of the stars in the

Milky Way are of this type. If they all have Earth-like planets, it increases our playing field enormously.' This is the same conclusion that Jon Swift came to, and this new discovery makes it even more feasible.

Another reason why red dwarf stars are interesting is that they live longer than the Sun. They are, on average, older and have had more time to develop stable life forms. Because they are less hot than the Sun, the habitable zone is also closer to the star. In the edible model of the solar system in my classroom, the Earth was a peppercorn 15 metres from the red cabbage that was the Sun. On the same scale, Kepler-186 f is also a peppercorn, but its star is a tennis ball and is only 6 metres away. For this reason, Kepler 186-f is referred to as a cousin rather than a twin sister of the Earth.

Quintana is cheerful when I speak to her, but sounds a little tired. She has been working very hard in the past few months. 'My interest in red dwarfs was originally aroused by Jill Tarter, who organized a workshop on these kinds of stars in 2005,' she tells me. 'She had invited astronomers, biologists and even geologists to discuss the habitability of planets orbiting red dwarfs. Since then, I've been fascinated by these small stars. Whenever we got a new set of data from Kepler, I would immediately look at them with a colleague to see if there was anything new. That's when we found Kepler-186 f.' To avoid a controversy like that around GJ 581 g, Quintana had to be sure that the planet she had found really was a planet, and not a background star or a stellar pulsation. 'The referees assessing our article sent it back 40 times before it was finally accepted for publication in *Science*,' she says.

And now? Can we learn any more about Kepler-186 f, the Earth's cousin? Can we ever determine that life is not only possible there, but actually walks (or crawls) around on its surface? The answer is no. The star is 500 light years from the Earth, too far away to be able to study its atmosphere with the current or the following generation of spectrographs. The starlight that shines along the edges of the planet is simply too faint. Nor is it possible to travel to Kepler-186 f. Seth Shostak, the SETI scientist and radio man who always has a joke at his fingertips, wrote in a press release, 'It's so far away that even if you booked a trip on the speediest of our rockets, you'd have 100 million years to polish your Sudoku skills en route to Kepler 186f.' Shostak did, however, immediately launch a SETI campaign to enable a signal from the planet to be received with radio-telescopes on Earth. Just in case E.T. tries to phone us.

Quintana is primarily fascinated by how Kepler-186 f was formed. In addition to the 'discovery article' in *Science*, she also described in a detailed study how a planetary system like Kepler-186, with a number of small planets very close to the star, can ever evolve. Planets are formed from the flat, spinning 'birth disc' of dust and gas that rotates around a young star like a pancake. Accretions of dust in the disc rotate and, like clumps of dust around the house, gather up more and more material until all that is left of the disc is planets. The more material there is, the more planets can be formed. As most of the material is in the outer zones of the disc, the majority of planets will also be formed there. That means that planets that are as close to their stars as Kepler-186 f probably ended up there at a later stage. 'We think that these guys were not formed where we now see them,' Quintana tells me. 'We have shown how they can form in the outer disc, far away from the star, and later migrate inwards.'

This study of how exoplanets are formed may seem like a sidetrack, but it is very relevant to answering the question of whether extraterrestrial life can exist. Living organisms on Earth are made up of molecular building blocks. Cell walls, enzymes and DNA are all compounds of smaller, complex molecules. It was long believed that these structures evolved only on Earth – in Darwin's 'warm little pond' – after the planet had sufficiently cooled. Recently, however, we discovered that such organic structures are also found in comets and meteorites. During and shortly after the process of planet formation these projectiles, originating from the outer regions of the solar system, fly around in all directions. Could the building blocks for life perhaps come from space?

nine

BEER IN SPACE

'SPACE IS a cold and barren place. Nothing can exist there, nothing!' Ludwig Von Drake, an obscure uncle of Donald Duck and a professor of astronomy, is sitting on a high stool in his observatory. When he sees that he is being filmed, he falls off and lands on the floor with a loud thump. 'Now I can see stars I've never seen before!' he groans. He walks over to a table with a large pile of books on it. The thickest of them all is a guide to space travel that he wrote himself. In a 45-minute-long monologue, he tells us in a thick German accent how mankind discovered the planets in our solar system and has fantasized about everything that might be crawling around on them. Every now and then, he picks up a book from the large pile and reads from it, and then throws it nonchalantly into a corner of the room. He tells us about Copernicus and Galileo, and about Kepler's dreams about Martians, Fontenelle's speculations about life on other planets, and even John Herschel's Great Moon Hoax. Science fiction comes to life in the colourful cartoon: hairy space beings and flying saucers shoot across the screen. At the end, the professor has the last word. He finds all these fantasies poppycock; nothing can live in that empty, barren space! But, as he is speaking, Von Drake is kidnapped by a black Martian robot from one of his stories.

The cartoon, *Inside Outer Space*, is part of *Walt Disney's Wonderful World of Color*, a television series from the 1960s. The absent-minded duck professor hosts a number of episodes, each with their own topic: the history of flight, the colour spectrum, space – all exciting stuff for American kids in the Space Age.

Lou Allamandola spent his teenage years in the science-crazy 1960s. He grew up in a Catholic family in the state of New Jersey. His grandparents were immigrants from Italy, and he didn't learn to speak English until he went to school. He still clearly remembers the Disney cartoons with Ludwig Von Drake, which were broadcast on Saturday evenings. 'Von Drake called the interstellar medium – the empty space between the stars and the planets – a barren place where nothing could exist,' he tells me. 'That was all we knew in the sixties. Now we know better. Interstellar space is full of molecules that we also see on Earth.'

I speak to Allamandola one Wednesday morning while he is visiting Leiden Observatory. He is a tall man with curly hair, greying at the temples. His accent reminds me a little of the Italian-American mafiosi from *The Sopranos*. While we are talking, the door of his office opens now and again, colleagues who urgently need his opinion on their latest research results or to correct an article they are writing together. He tells them all to come back in the afternoon. 'When I'm here, far away from my own office and telephone, I find it easier to say no,' he says. That office is at NASA's Ames Research Center in California. Since 1983, Allamandola has been head of the Astrochemistry Laboratory, where they study how molecules behave in simulated space conditions. Astrochemistry – space chemistry – is a relatively new discipline, and Allamandola is a pioneer in the field.

ON 20 July 1969, the peak of the Space Age, hundreds of millions of people sat glued to their televisions and radios, following *Apollo 11*'s Moon landing. They heard Neil Armstrong's words above the background interference: 'That's one small step for [a] man, one giant leap for mankind.'[26]

It is remarkable just how little we knew back then about the chemical composition of the interstellar space through which the astronauts were floating. Admittedly, compared to the Earth, space is very empty. The lowest density that can be achieved with the best vacuum pump on Earth is still a trillion times higher than it is in space. That is roughly equivalent to the difference in density between air and a stone wall.

And yet, we knew that space wasn't completely empty. At the beginning of the twentieth century, telescope photos of areas with many stars showed strange dark spots where there were no stars at all. They proved to be large clouds of gas and cold space dust that absorb the light of the stars behind them. What lay hidden inside these dark clouds, however, could be made visible using spectroscopy.

Every atom can absorb and emit at specific wavelengths, resulting in a fixed pattern of absorption and emission lines on the spectrum. This 'fingerprint' can be measured with a spectrograph. Mayor and Marcy measured the changes in wavelength of these lines in stellar spectra, so that they could use the Doppler method to determine the speed at which the stars were moving.

It is not only individual atoms that have spectral lines. Molecules – combinations of atoms – also emit light at certain wavelengths. These wavelengths are determined by the movements of the molecule. Hydrogen, the simplest molecule, comprises two hydrogen atoms that are joined together. That combination is possible because the two atoms share their two electrons. You can see it as two small balls connected by an elastic band (the electrons). Because the elastic is flexible, the atoms can move back and forth, like a kind of stretching exercise. And they can do that at varying speeds. If they change their speed or direction, they emit a light particle. These particles, called photons, each have a specific wavelength. That means that the light emitted by a gas cloud in space contains the spectral lines – the fingerprint – of the molecules within the cloud. In short, we can tell from the light from a gas cloud what kind of molecules it contains.

Molecules were not first detected in space until the mid-twentieth century. This had not been possible earlier because their spectral lines have very long wavelengths and can only be detected with radio and infrared telescopes. In 1800 William Herschel was the first to detect infrared radiation from space, but it would take a long time for better instruments to be developed.[27]

Radio astronomy, too, did not really pick up momentum until the 1960s, thanks to technology developed during the Second World War. Frank Drake and his companions used it for their early SETI experiments, but astronomers interested in star formation also studied radio waves. Gas and dust clouds were mainly found in the middle of groups of young stars, suggesting that stars were born in the clouds. As the cloud cools off, the particles within it move more slowly until the cloud collapses under its own gravity. The material in the middle of the cloud then condenses and a new star forms. Astronomers hoped to learn more about this process of formation by studying the radio spectral lines from the birth clouds.

The first molecules found in interstellar gas and dust clouds using radio observations had a very simple composition, with no more than two atoms per molecule.[28] In March 1969 the discovery of the most complex molecule to date was announced: formaldehyde, which has the chemical formula CH_2O. The article announcing the find, the lead author of which was radio astronomer Lewis Snyder, closed with the words that 'molecules containing at least two atoms other than hydrogen can form in the interstellar medium.'

A certain degree of surprise can be detected in this statement: it had been assumed until then that there was nothing to be found in space. It was Ludwig Von Drake's 'barren place', a godforsaken void where no molecule can survive. And now experiments were being conducted that suggested that the space between the stars was teeming with complex chemical matter.

Snyder's paper appeared four months before the Moon landing, making the contrast even greater. Mankind could send astronauts into space, but had no idea of the chemical riches it contained.

ALLAMANDOLA LAUGHS and shakes his head when he thinks of all the many discoveries that were still awaiting astronomers. In 1968, he graduated in chemistry from St Peter's College, a small Catholic university in New Jersey. 'Miraculously enough,' as he says himself, he was selected to conduct PhD research at the prestigious Berkeley, which had one of the best chemistry departments in the country. His mentor was chemist George Pimentel, 'a wonderful man, who possessed the skills of ten people', says Allamandola. One of the many interests of the multifaceted Pimentel, who also invented the chemical laser, was measuring the infrared spectra of gases in his laboratory. He wanted to apply this technology to find out whether there was life on Mars by detecting gases originating from life forms. NASA sent a spectrograph he had built himself on the unmanned Mariner spacecraft that flew past the red planet. The spectrograph did not discover any biological material, but did provide a lot of information on the temperature and conditions on the planet's surface. Following this, NASA selected Pimentel to be part of the first group of scientists to be trained as astronauts. He withdrew from the programme, however, when it became clear that he would probably never go into space.

While studying under Pimentel, Lou Allamandola learned about infrared spectroscopy in the lab. After getting his PhD, he found a research position in Oregon, a little further up the west coast of the U.S. When his contract expired in 1976, it was difficult to find a new job. 'The oil crisis had hit, and there was little money available for research,' he explains. 'Instead of the four or five offers I would normally have got ten years previously, I had about 80 rejections. My wife and I had just had our second child and we were a little at a loss about our future. Then I had a call from George Pimentel. He had heard about a position that was made for me. An acquaintance of his, theoretical astronomer Mayo Greenberg, wanted to set up a lab to simulate chemical processes in interstellar dust clouds. That was music to my ears. Then George said "Just one snag. How is your Dutch?"'

During his subsequent telephone calls with Greenberg, Allamandola became more and more enthusiastic about the work he would be doing in Greenberg's lab in Leiden. Until then, astronomers had mainly found space dust irritating, as dark dust clouds obstructed their view of areas where stars were being formed. But Greenberg found them fascinating. He suspected that grains of space dust were covered in an outer layer of water ice, like snowballs, in which other chemicals – for example oxygen and carbon – were dissolved. Allamandola explains how Greenberg came to this conclusion: 'Space dust contains the element silicon, just like glass. Water vapour drifting around in space condenses onto the silicon in the same way that, here on Earth, we see ice flowers on our windows when it freezes outside. The glass cools the air and the vapour in the air freezes. It ain't magic, but for some reason or another, the snowballs had not yet occurred to most astronomers.'

Greenberg and Allamandola became interested in the frozen dust grains because all kinds of chemical processes could occur in them that were impossible elsewhere in space. 'Imagine a lonesome molecule drifting through the vacuum of space,' Allamandola explains. 'After a few hundred million years, it comes across another molecule with which it reacts and forms a new molecule. That process will be speeded up if the molecules are more closely packed together in ice that has settled on space dust.'

The ice, which – compared to interstellar space – has a very high density, acts as a kind of meeting place for molecules. When the surface of a dust grain is illuminated by a star, it activates all kinds of chemical processes. The energy supplied by the ultraviolet starlight enables larger molecules to be formed from small chemical building blocks.[29] If Greenberg's suspicions were to prove correct, a wide range of complex molecules could be formed in the interstellar ice grains. Perhaps the chemical materials to create organisms on Earth originally came from space.

And so, in 1976, Allamandola and his young family moved to Leiden. He was to stay there for eight years, and he says his Dutch is 'still reasonable'. He shows me a picture of the research team at the Leiden laboratory in 1970s. Eight men and one woman. They have long hair, black-rimmed glasses, some of the men have thick beards. Greenberg himself is in front of the group: a small man with grey hair, a blue roll-neck sweater and a tweed jacket. The laboratory assistants are surrounded by complex-looking equipment.

Allamandola tells me that doing research in the 1970s was very different from now. 'We didn't have these things,' he says, tapping his laptop screen. 'It used to be normal to talk to each other for hours in the canteen. About science. If you wanted to read an article, you went to the library, where you could spend the afternoon thinking in complete peace and quiet. I don't know many people now who still spend an afternoon sitting and reading. There is always the pressure to do so many other things. At conferences, people check their emails rather than listen to the speaker. On your laptop, you have the entire canon of scientific literature at your fingertips, but that doesn't mean you absorb the information more quickly. There are Schwarzenegger movies about how machines will take over the world. In my opinion, in a certain way, they already have.'

Allamandola shows me the next photograph, a close-up of the machine that the researchers were standing around. 'Look, this is an ice simulation chamber. Normally, I don't like having to explain complicated measuring equipment, but this is so simple. It exactly recreates the situation in space that we want to replicate.' Without an explanation, the machine indeed looks complicated, a little like the inside of a computer. There is a lamp aimed at a kind of biscuit tin with a tube screwed to it. 'This emitted ultraviolet light and simulated the star,' Allamandola says, pointing to the lamp. 'The biscuit tin represented the dust cloud; inside it was a deeply cooled sample of water ice containing ammonia and carbon monoxide, two common molecules in space. The tube behind it was the spectrograph. That captured the light, registered whether molecules had been formed in the ice and, if so, which ones.'

It worked. Allamandola shows me two spectra – one before and the other two hours after radiation with ultraviolet light. The first spectrum showed only lines of water, carbon monoxide and ammonia – the ingredients of the ice sample. But the second contained a large number of new spectral lines, indicating the presence of new, larger molecules that had been formed from the basic ingredients.

This result was spectacular. In the vicinity of stars, the ice mantle of the space dust grains became molecule factories, in which a wide range of complex structures could be produced. In 1969, scientists had been surprised to discover that a complicated molecule like formaldehyde could be formed in space. Yet in the ice chambers in Leiden, under the same conditions as in space, it was produced in great quantities from the 1970s onwards.

But the results of the laboratory experiments were not immediately noticed and accepted by others. 'Astrochemistry was still a young discipline,' Allamandola tells me. 'Scientists discovered more and more new molecules in space. They constructed theoretical models that show how the molecules – in gas form, not in an ice crystal – could have formed. The fact that these reactions could never take place if the molecules were just drifting separately through space was ignored. Astrochemists were just fine without our icy grains. They saw us as nutty professors.'

That all changed in the 1980s when Allamandola and his colleagues, including Leiden astronomer Xander Tielens, conducted observations on board the Kuiper Airborne Observatory – a converted Lockheed aircraft fitted with a telescope and spectrograph. The telescope was behind a hatch in the side of the fuselage. An airlock made sure that the researchers were not sucked out of the plane as a result of the fall in cabin pressure when the hatch was opened. Because the plane could climb above the water vapour layer in the atmosphere, it was possible to measure the water vapour and ice in space. And they found the ice grains: dust cloud from which stars and planets were formed contained water ice and the same complex molecules that had been produced in the laboratories in Leiden and Ames.

DURING A conference in Australia in 2010, I myself first heard about the many molecules that have been found in interstellar space since that time. The conference dinner was on Magnetic Island, off the east coast of Queensland. On the restaurant lawn, possums scratched around between the laid-out tables. Some two hundred astronomers had just finished their dessert and Andrew Walsh, the conference organizer, was speaking. Walsh is a slightly curt Australian with little hair on his head and a beard in two impressive long plaits. Besides astronomy, his great love is brewing beer.

'My father asked me when I started doing my PhD in astronomy "So what do you actually spend your days doing?"', Walsh told us. 'I told him the title of my dissertation: "The Association of Ultracompact HII Regions and Methanol Maser Emission". He looked at me glassy-eyed, and I could see that I was losing his attention – until I said the word "methanol". "Aha!" he said, "so you have alcohol in space? Is there beer in space?" I explained to him that ethanol, not methanol, is the alcohol you get in beer. "Methanol is a poison, Dad," I said. "If you drink just a little of it, you'll go blind. If you drink any more, you'll die." From that moment, my father lost all interest in my work. I'd like to put that right with this presentation, which I give the title "Beer in Space" and dedicate to my father.'

In fifteen minutes, Walsh – who became more and more animated – listed the twelve main ingredients of beer. Water, alcohol (ethanol), sugars, a few amino acids. Then he showed us photographs of the areas where stars are formed – the same dust clouds that Allamandola simulated with his laboratory ice. Enthusiastically, one after the other, Walsh named the ingredients of beer that have been found in all these clouds: plenty of water and ethanol, carbon dioxide, even sugars and a few simple amino acids. Five of the more complex amino acids and sugars have not yet been found, but Walsh is convinced that that is because we haven't looked hard enough yet. He called on his colleagues to keep looking for the missing ingredients for the space beer. 'It would be very reassuring for my father and many others to hear that we also find useful things in space,' he concludes.

FROM THE 1980s, astronomers had not only found some of the ingredients for beer in space, but had started a tentative search for the basic materials of life. Lou Allamandola returned to the U.S. in 1983, where he set up his own laboratory at Ames to continue the experiments he had conducted in Leiden. 'The list of substances we made in the lab got so long that even chemists were starting to find it boring. At the end of the eighties, we wanted to see whether we could also make molecules that resembled the building blocks of living organisms.' I ask Allamandola whether, as a religious man, it is difficult to combine his faith with a study of the origins of life. 'Not at all,' he says. 'Religion and science are different domains – and both contain great mysteries. Besides, the chemistry I study is still so far removed from the origins of life.'

Some of the experiments conducted by the Allamandola team produced remarkable results. After each experiment, the radiated ice was thawed out and dissolved in water. The liquid was then heated up so that the water evaporated. What was left was an oily residue that Mayo Greenberg had already christened 'the yellow stuff' in his early experiments. Perhaps there was something in the yellow stuff that was too complex to have been picked up earlier by the spectroscope? Greenberg made headlines in the Netherlands in 1980, when he suspected that the residue also contained amino acids. Amino acids are the basis of proteins in our bodies and are the building blocks of life. 'ORIGIN OF COSMIC LIFE SIMULATED IN THE LAB' was perhaps the least dramatic headline. The local *Leidse Courant* had no such reservations, stating with wild exaggeration 'LEIDEN RESEARCHERS FIND LIFE AMONG THE STARS'.

'Of course we hadn't made living organisms,' says Allamandola. 'You always have to watch what you say, otherwise people get the wrong idea. Prebiotic, biogenic . . . in other words, the same building blocks from which life is made. A human being, even a living cell, is an enormously complex Lego construction. All we found were a few individual Lego bricks, not the whole structure.' But they did find an enormous variety of chemical building blocks under the microscope. Besides amino acids, there were also sugars and even nucleic acids, which form the basis of DNA. They also found elongated molecules that repelled water on one side (hydrophobic) and bond easily with water on the other (hydrophylic). The cell walls of the human body are made of the same type of molecules.

As Allamandola tells me all this, I become as enthusiastic as the journalist from the *Leidse Courant*. They'd discovered that life in space is possible! Allamandola spreads his arms and gestures me to calm down. 'Ho ho, Lucas,' he says. 'No one knows what life is. There are about five hundred different definitions. What we have found has nothing to do with life, as yet. All we have found is the building blocks; how they eventually lead to a living organism is a completely different matter.'

SCIENTISTS HAVE been addressing this question for hundreds of years. In the 1950s Miller and Urey conducted experiments to explore Darwin's idea that life on Earth had been formed in a warm little pond that was struck by lightning. Complex molecules like amino acids were produced in their test environment, which was later more or less replicated by Bill Borucki. The experiments by Allamandola and Greenberg showed that the same substances could be created in a block of ice in space that had been radiated by a star. The big question was, how did these substances end up on Earth?

The Earth most likely originated as a hot ball of liquid rock. Some four billion years ago, it had sufficiently cooled for life to evolve. The oldest fossils found on Earth are of bacteria that developed around that time. The ice experiments showed that we could also find the basic materials for these organisms in space. Could these molecules, via a sort of cosmic postal service, have been delivered to the Earth after it cooled off? Panspermia, the hypothesis that life on Earth originated in space, was starting to look like an interesting possibility.

In 1989 Allamandola met biochemist David Deamer. At that time, Deamer had a fragment from a meteorite that had impacted in Australia. An enormous chunk of rock weighing around 100 kilograms had broken up into smaller pieces in the atmosphere. The fragments were later analysed in a laboratory. Deamer's meteorite proved to have the same cell-wall-like structure that Allamandola had created in his lab. It was a remarkable find, showing that meteorites that impact the Earth contain the basic materials for organisms. The time was not yet right, however, to draw far-reaching conclusions. 'There are still people who leave the room if they hear the word biomarker – an indicator of life – in a talk. I simply did not dare to show some of our results, which suggested that the building blocks of life can be formed in meteorites. If I were to do that, whether it was at a chemistry or an astronomy conference, my colleagues would think I had gone insane.'

In the mid-1990s, however, astrobiology became increasingly popular. In 1996, Allamandola was the speaker at a workshop organized by NASA and SETI on the island of Capri, off the west coast of Italy. At the end of his presentation, he dared show a slide depicting the structures of Deamer's meteorite alongside those from his own laboratory. 'The time was ripe,' he tells me. 'People were open to the idea that organic materials on Earth may have been delivered by meteorites.'

SINCE THEN, there has been a growing awareness that many of the substances that we absorb on a daily basis were formed in space. Take water, for example. Every meteorite or comet is an enormous snowball with its origins in the birth cloud of the solar system. If such an object impacts the Earth, it deposits a large quantity of water onto the planet's surface. It is difficult to imagine that enough of these snowballs have fallen to Earth to create all the oceans, but I recently saw an image that made this idea a little more acceptable.[30] It was a drawing of a dried-up Earth, with the water from all the rivers, oceans, lakes and so on condensed into three small spheres. The largest of the spheres – in diameter about the same as the distance from Amsterdam to Rome – represents all the water in, on or above the Earth. It is pretty small compared to the size of the Earth. Suddenly, it didn't seem so strange after all that every glass of water, every cup of tea and every beer that I had ever drunk had once been part of a snowball in space.

A meteorite impact may not seem like something that happens every day, but it is. Only the larger impacts make the news, but thousands of kilograms of interstellar material land on the Earth daily, in the form of small meteorites and space dust. In the young solar system, the impacts were even more frequent and violent. Dating of craters on the Moon shows that around four billion years ago, enormous showers of meteors passed through the solar system for a period of millions of years. These must have impacted on the Earth as well as the Moon.

One possible explanation for these meteorite showers is that, shortly after it was formed, the planet Jupiter moved a little closer to the Sun. This migration was apparently caused by the gravity of other planets and small objects orbiting the Sun. The shift in Jupiter's orbit must have knocked the rest of the solar system out of balance and acted as a sort of catapult on the space rubble flying around the planets. Consequently, the inner planets – including the Earth – were severely bombarded with meteorites for a long period. This event has come to be known as the Late Heavy Bombardment. Similar bombardments are also observed today around young stars that are still in the process of formation. Space dust and water are thrown around by the embryonic planetary system and end up on the planets as they cool off.

One of the most well-known images from the Hubble Space Telescope has been nicknamed 'The Eye of Sauron' by astronomers, because of its close resemblance to the symbol of the Dark Lord in the *Lord of the Rings* movies. The image shows a kind of golden halo, surrounded by an oval ring. In the centre of the ring, the star has been removed because it is too bright. This has left an elongated dark spot on the image that looks like the pupil of an eye.

The image is of Formalhaut, one of the closest stars to the Earth. The oval is reflected light from a ring of space dust. The dust is probably produced by comets and other space rubble flying through the system at random. Every day, thousands of objects collide, break into smaller pieces and produce space dust that is full of water and organic molecules. Large and smaller fragments eventually land on the young planets orbiting the young star. The comet shower in Formalhaut shows us what the Late Heavy Bombardment probably looked like.

We are currently finding out a lot more about these water-carrying projectiles in our own solar system. In 2014, the spacecraft *Rosetta* reached comet 67P/Churyumov-Gerasimenko. It sent down the lander *Philae*, while the mother ship continued to orbit the comet for two years before crashing intentionally on its surface. *Rosetta* and *Philae* found water, oxygen and numerous organic compounds (not to be confused with living organisms) in the comet. Curiously, the molecular make-up of the comet's water was very different from the water on Earth, suggesting that comets – or at least, comets like 67P – might not have contributed greatly to the delivery of water to Earth. The eventful Rosetta mission marked the first time in history that comet water and dust could be studied directly.

WHEN I had finished talking to Allamandola, I felt as though I myself had undertaken a cosmic journey. In the two hours we spent together in his office in Leiden, we followed the passage of an organic molecule through space; from its formation in a frozen grain of dust in the birth cloud of a young star, via the disc of dust and gas in which planets are formed, to its arrival on a planet through a meteorite impact.

It is a route that is still the subject of intense study by astronomers, including those in the Netherlands. Allamandola is in Leiden to give lectures at the two leading astrochemical research groups located there, one of which is led by his friend and former colleague Xander Tielens. Telescopes like the infrared satellite Herschel and ALMA, an array of several dozen radio dishes in the Chilean Andes, are exposing parts of the spectrum that were previously not accessible. This produces new spectral lines and new molecules in star formation regions.

These observations make some planet hunters optimistic about the chances of life existing on exoplanets. After all, the materials from which the inhabitants of Earth are made are also found in young planetary systems. Space is not the barren, empty place that Ludwig Von Drake described, but is teeming with the building blocks of organic life. Dissolved in water, these materials are continually delivered to the surface of young planets by meteorites. If the temperature is right and the ingredients are present, time and evolution do the rest. Perhaps it was this line of thought that led planet hunter Steven Vogt to make his '100 per cent' claim about Zarmina's World.

Yet just how the route runs from building blocks, via chemical reactions, to life remains unclear. We do not even know how it happened on Earth. Direct evidence – for example, the earliest life forms – have as far as we know largely disappeared from the face of the Earth. There are too many uncertainties to give preference to any single theory on the origins of life above all the others. And, for that reason, we cannot use life on Earth as a blueprint for the rest of the universe. Most planet hunters take a different approach to the question of whether *extraterrestrial* life exists. Imagine that a certain life form had developed on another planet from the same building blocks that we use on Earth and that we see everywhere in space. How then could we detect the existence of that life form from the Earth? How can we recognize a sign of life from an exoplanet?

ten

THE SPACE REBELS

AFTER A rare lull in the conversation, Sara Seager says: 'Do you know what the best thing about this place is? If you have an idea, no one says "That's crazy." The answer is always "Let's do it."' Seager, professor at the Massachusetts Institute of Technology – better known as MIT – looks out at the spectacular view from her office on the seventeenth floor. The River Charles stretches out from left to right and dozens of small sailing boats bob up and down on the water in the late afternoon sun. On the other side, the serrated skyline of Boston's skyscrapers is sharply defined against the horizon. 'People who work here want to cross frontiers,' Seager says. 'It's in the air.'

The day starts 5 kilometres away, at the Center for Astrophysics, the stately brick astronomy building of Harvard University. Near the gleaming coffee urn, a small group of planet hunters recall memories of the early days, in the 1990s and earlier, before their hobby got out of hand. After that, I leave for MIT. It didn't seem very far on the map, but appearances can be deceptive. First of all, it takes at least three-quarters of an hour to cross the city in the busy traffic, giving my taxi chauffeur ample time to tell me all about his experiences as a paratrooper in the jungles of Mozambique. What's more, once I see the white tower blocks looming up in front of me, I feel as though I've arrived in a different world. Harvard is built of brick, MIT of concrete. Harvard is the establishment, MIT are the rebels with new ideas and controversial methods. The taxi ride feels like a journey through time.

I get out at the pillar-lined entrance and call Sara Seager's office. On the other end of the line, I hear the dry voice of Derrick, Seager's personal assistant. This is the first time I have had dealings with an astronomer's personal assistant. As far as I know, most professors fill in their own expense statements. One of Derrick's jobs, he tells me, is to screen telephone calls. Seager gets calls now and again from people who claim to have seen UFOs. Derrick does not put them through to his boss. After five minutes, Derrick appears at the entrance. I was expecting a typical nerd but, instead, see a small, muscular man with dreadlocks. 'Hi, I'm Derrick,' he says. 'Let's go into the infinite corridor.' I imagine myself in *The Matrix* and follow Derrick up the stairs to the main entrance.

The 251-metre-long, straight corridor seems to have no end. It is the main artery of the institute, connecting five different main buildings. Students hurry past me to the left and right. On the wall, I see posters for conferences, prototypes of new inventions, awards and busts of university legends. We walk through and past reading rooms, canteens, classrooms, stairs and lifts. Although there is nothing here you can't find at other universities, it feels somehow different. Exciting, and innovative. At each classroom, I try to see what the lesson is about. Nanotechnology? Egyptian dung beetles? There is an incessant flow of students in the infinite corridor, cheerful and full of energy, and the countless side passages are filled with their chatter, too. I have arrived in the organized chaos of an anthill. By the time we get to the other end of the corridor, I am completely star-struck. This is a special place. Innovation and ambition hang in the air. You can taste them.

We walk a hundred metres or so across a courtyard to the highest building on the campus. On the seventeenth floor, Sara Seager greets us. She is a slim woman in her early forties with sunglasses in her black hair. I have already heard a lot about her, that she was ambitious, intense, 'out there'. She is extremely alert and speaks quickly. Very quickly. Because it is all so interesting, I sit on the edge of my chair and give her my full attention. I talk to her for three-quarters of an hour, but it seems more like three hours. Not because it is dull, but because she covers so many subjects. The cogs in her mind seem to turn exceptionally quickly; the conversation goes off in all directions, but she never seems to lose control. She is open and interested in what I have to say. She asks continually about my own background and why on earth I am writing a book, and tells me about her own successful and failed projects.

As we look out of the window together, she says 'Boston is a wonderful city to live in. You can go to the ocean or the mountains. They're not actually real mountains of course, but they're close enough.' The rather civilized east coast of the u.s. is completely different to the Canadian wilderness that she travelled through in her student years. Her old student friend David Charbonneau had told me about her love of adventurous trips. With her husband Mike, Seager would sometimes disappear for months at a time, canoeing up a mountain river, only to eventually resurface at some remote railway station.

Her life took a dramatic turn two years earlier, when Mike died unexpectedly. She was suddenly alone, with her two young sons. She wrote in a blog:

> I have a big problem on my hands. Not only dealing with losing my husband, but on how to do anything around the house. One of my secrets to success is that in the early years of my career I could focus on work because Mike took care of absolutely everything on the home front. In 2010, I didn't know that at the grocery store you had to swipe the credit card. I hadn't put gas in the car for as long as I could remember. I almost never visited the children's school.

After a tough time, Seager crept slowly back out of her dip. The people she worked with were a great help. 'The students and researchers became my extended family,' she tells me. 'We go on holiday together. The kids love them. My house is always full of people.' A week after our meeting, she receives a MacArthur Fellowship, a very prestigious research prize – also known as a 'Genius Grant' – awarded annually by the John D. and Catherine T. MacArthur Foundation. She spends the money entirely on childcare and household help. 'I'd rather think than scrub,' she says in an interview.

Her clear formulation and straight talk make Seager the ideal spokesperson for science. She is often called upon to state her case in political debates on the national budget for the search for extraterrestrial life. In 2013, for example, at a hearing on the subject of 'space aliens' in the House of Representatives, Seager said that, if extraterrestrial life exists, discovering it is only a matter of time and money. Passionately, she encouraged the politicians to support the search, telling them 'This is the first time in human history that we have the technological reach to cross the great threshold.' Even a Republican representative known for his attacks on the science budget was impressed, saying: 'I don't know how I'm going to tell my barber or the folks in my home town about your testimony, but you must clearly enjoy getting up in the morning and going to work again.'

Seager describes her career in science as a 'random walk' through a maze: choosing a direction at random every time she comes to a crossroads. She has been working on the design of the maze since primary school, where she spent most of her time daydreaming. 'I still follow the winding paths I laid down in my mind back then when I conceive and evaluate new ideas,' she says. Her parents encouraged her not to conform: 'My father taught how me to think big. He introduced me to many outlandish ideas at a very early age, which helped me to take seemingly crazy ideas seriously, even my own.'

She developed a natural distrust of authority, and the guts to doubt everything that she didn't understand or believe. That gave her an adventurous desire to try new things. 'I always took risks,' she says. 'And I still want to do what no one else does. Then you don't feel any competition, that unpleasant sensation of people breathing down your neck.' I recognize that feeling and think back to 1993, when the film *Jurassic Park* came out. Half the school was suddenly showing off their dinosaur collection in the school yard and I lost my fascination with prehistoric animals forever; it was just no fun anymore.

Seager's interest in astronomy was aroused in a familiar way – gazing at the stars during her first camping trip. She immediately saw the cosmos as a good place to take her 'random walks'. Her lust for adventure is also reflected in her scientific work. 'Most science is business as usual,' she tells me. 'Here, we try and do things that are exactly not that. I am engaged in projects that have a substantial risk of failure, but a very high return if they succeed. That is my privilege and my obligation here at MIT.' She also tries to encourage her students to take on unconventional projects from time to time: 'A large part of their research is the bread-and-butter stuff, work that is guaranteed to produce publishable results. As a student, there's no getting round that, but the risky things are often more fun. So I encourage them to devote part of their time to them, too. It stretches their minds more.'

The kinds of projects Seager is talking about are also known as 'Friday-afternoon experiments', crazy ideas with little chance of succeeding, but with enormous potential if they do; experiments conducted while the rest of the institute is already in the bar getting the weekend off to a good start. Mostly, they result only in social isolation and a few missed beers, but some accidental discoveries in lost hours have led to great breakthroughs. In 1922, while suffering from a bad cold, biologist Alexander Fleming dripped the contents of his own nose onto a Petri dish. After a week he saw that, although the dish was crawling with bacteria, there were none at all around the nasal mucus. This observation eventually led to the development of penicillin.

Russian-Dutch-British physicist André Geim elevated Friday-afternoon experiments to a tradition at the universities of Nijmegen and Manchester. In 2000, he won the Ig Nobel Prize (the alternative Nobel Prize, especially for Friday-afternoon experiments) for levitating a frog in a magnetic field. Ten years later, he was awarded a real Nobel Prize for an equally improbable breakthrough. He had discovered graphene, a super-thin but immensely strong form of carbon, on a piece of sticky tape that he had removed from a pencil. Graphene is expected to have a wide variety of applications in the future in, for example, computers and smartphones. In an interview, he said that he values both his prizes equally, as 'a little bit of self-deprecation always helps'.

A Friday-afternoon experiment that got out of hand early in Sara Seager's career resulted in – in her own words – a gigantic failure that eventually had a silver lining. In 2000, while doing her PhD research at Harvard, she devised a method, together with Dimitar Sasselov, to detect the atmosphere of exoplanets. When a planet passes in front of a star, a small amount of the star's light will pass though the planet's atmosphere on the outer edge of its silhouette. Absorption lines can be detected in this light, which will tell us what chemicals are to be found in the atmosphere. This is the method David Charbonneau and Tim Brown used successfully in 2002 on HD 209458 b, the planet they observed from their yellow hut.

Seager wanted to find more planets with visible transits and use the method on them, too. By then, she had been taken on by the Institute for Advanced Study in Princeton, the renowned research institute where Einstein spent the last twenty years of his academic life. According to its mission statement, the Institute encourages 'the curiosity-driven pursuit of knowledge with no view to its immediate utility or the expectation of meeting predetermined goals'. Academics are given complete freedom to determine their own research agenda.

At Princeton, Seager – who had until then mainly focused her skills on writing computer programs ('I could just code for twelve hours a day without making a mistake') – became friends with Gabriela Mallén-Ornelas, a Chilean astronomer with extensive observation experience. Together they initiated a campaign to search for planetary transits, using an American telescope in the Atacama Desert in Chile. Because of its dry climate, the desert is the perfect place to look at stars: there are no clouds or water vapour in the air to distort the images and many international telescopes are located there. Chile was also an attractive prospect because, as a Chilean national, Mallén-Ornelas was entitled to more observation time.

Unfortunately, the observation campaign was literally a washout. The unpredictable El Niño had struck and the normally dry weather had been brought to an abrupt end. During most of the time allotted to Mallén-Ornelas and Seager, the observatory was ravaged by snowstorms; they could hardly have been more disappointed.[31]

In the observations they were able to carry out, they found no transits, largely because they had not sufficiently thought the planning through. So as to still learn something from the project, Seager and Mallén-Ornelas spent one hot summer evening talking about everything that had gone wrong and what they could have calculated if they had been able to observe a transit. They started studying the mathematical formulas that described the movement of a planet as it passed in front of its star. 'Around midnight, we both suddenly had a Eureka moment,' Seager tells me. 'We shouted "Density!" in unison and burst out laughing.' They had both realized at the same time that, by combining the right formulas, they could calculate the density of a star directly from the depth and duration of a transit. It is useful to know the density of a star, as it enables a number of properties of the star and its planet – such as their absolute size – to be defined.

Some months after Seager told me this Eureka anecdote, it was put into perspective by a Dutch professor. He pointed out to me that the formula that Seager and Mallén-Ornelas had 'discovered' had already been in existence for several decades and was used to calculate the densities of binary stars. Either way, the article they wrote about it in 2003 attracted a lot of interest and is still Seager's most quoted exoplanet article. A failed observation project that produced a significant theoretical result – that's a Friday-afternoon project to be reckoned with.

SEAGER'S ACTIVITIES at Princeton were followed at a distance by Professor John Bahcall, who founded the astronomy group at the Institute. From the 1970s on, Bahcall had been one of the driving forces behind the development of the Hubble Space Telescope. Shortly after Hubble was launched in 1990, a mirror proved to be malfunctioning and a manned mission was needed to repair it. Members of Congress were not keen to send astronauts into space to do the job, partly because the makers of the telescope had, as yet, not been able to tell them exactly what it was going to discover.

It is of course always impossible to predict unexpected and revolutionary discoveries, but try and get government support with an argument like that. Bahcall was invited to attend a hearing of the House of Representatives and passionately defended Hubble, concluding that 'the most important discoveries will provide answers to questions that we do not yet know how to ask and will concern objects that we cannot yet imagine.' His words and the hefty report accompanying it convinced the politicians; the Hubble telescope was repaired and Bahcall's prediction would be fulfilled many times over.[32]

At Princeton, too, Bahcall continued to give his passionate support to unusual research. 'John was very enthusiastic about our project to observe new planetary transits,' Seager tells me.

'I don't know if he really believed that two inexperienced people like us would succeed. But he still asked us every morning how the observations in Chile were going. I think that he mainly wanted to encourage me to try new things. All he said to me when the project failed because of the bad weather and an error caused by inexperience was "Sara, you need to understand data."'

That comment was an important lesson for Seager about the scientific method: you can't build houses if you've never stood with your feet in the mud. A theoretical astronomer also has to know exactly what a telescope can and can't do. A theory has to be proven by observation, otherwise it has no value at all.

The observation programme in Chile with the unexpected outcome illustrates a recognizable pattern: Seager signs up on a new ship, takes over the helm and finally jumps into a lifeboat, in which she paddles on her own to an undiscovered island. That was what she did when, after three years at Princeton, she had to look for a new job.

In 2002 she moved to the Carnegie Institution for Science in Washington, DC. Like Princeton, this institute is well known for its groundbreaking and interdisciplinary research. It was here that Edwin Hubble published his observations on the expanding universe, botanist Barbara McClintock explained how mutations occur in genetic material, and geophysicist Charles Richter devised his system for measuring earthquakes. It was one of the first places in the U.S. where scientists researching exoplanets established a foothold. 'I had applied to so many places,' Seager says. 'Our prediction on detecting exoplanet atmospheres had not yet been fulfilled, and people at my presentations were often derisive about it, claiming that we would never succeed. But there was always some older professor in the front row who would come up to me afterwards and say something to the tune of "If I were young, I'd study exoplanets, too."'

At the Carnegie Institution, Seager found herself in the Department of Terrestrial Magnetism. The department had been set up a century earlier, with the aim to 'understand the physical Earth and the universe that is our home'. From the 1970s onward, that mission acquired a new dimension, with the discovery of planets outside the solar system. It suddenly became conceivable that, in the future, we would find planets that closely resembled the Earth. Besides planet hunters like Paul Butler and planetary theorists like Alan Boss, there were a lot of geophysicists working at the Carnegie Institution. Geophysicists study the composition of the Earth and the processes occurring in its interior. Seager familiarized herself with this field of research and applied it to the as yet undiscovered Earth-like exoplanets. In 2007, she devised a formula that can be used to calculate the mass of an Earth-like planet if only its size is known (and vice versa). She also further refined the definition of super-Earths, the first examples of which had been found around that time.

At MIT, where the desire to innovate was embedded in the foundations, they were crying out for planet hunters. In 2007 they created a professorship specially for Seager and told her she could set up her own research group. Her father, whose health was failing rapidly, advised her to take the position as such opportunities were rare. But when she told him that she had taken the job and that a professorship at 35 was the best she could have achieved at her age, his eyes flashed back at her as he said 'I never want to hear you say that anything is the "best" you can do. I never want you to be limited by your own negative thinking.' It was one of their last conversations. 'He decided to finish by pushing me to keep thinking big,' Seager later recalled.

WITH HER father's words in her mind, Seager concentrates on answering one big question: every day, she devotes all her time and energy to finding extraterrestrial life on exoplanets. To achieve that ambition, she is working on three big projects at the same time, each of which represents an essential step on the road to discovering life. The first step is to find a second Earth (which goes without saying), the second is to observe its atmosphere (to determine which gases are produced on its surface) and the third is to identify a sign of life in this composition of gases.

Finding a habitable planet is at the top of the agenda. 'Around what kinds of stars can we best find and study Earth-like planets?' Seager asks. 'Many people think that red dwarfs are the most promising. But I think differently about that.' In her opinion, a second Earth must preferably be found around a close, bright star. A star that is no more than tens of light years from the Earth. Not necessarily because that will make it easier to go there, but because closer stars are generally the brightest.

The brighter the star, the more accurately the light it emits can be measured. Even Earth-like planets – which, because of their small size, only cause very subtle changes in light ('shallow' transits) – can be detected with these bright stars. And bright stars have clearer absorption lines in their spectrum, making it easier to confirm the subtle effect of a smaller planet using the Doppler method. It is also easier to observe the planet's atmosphere, because the light shining through it is brighter and the absorption lines are more clearly defined. This would be impossible with the small planets that Kepler has found, like Kepler-186 f, because these systems are much too far away. Seager's conclusion, therefore, is that we will find the most useful Earth-like planets close to home.

The Transiting Exoplanet Survey Satellite (TESS for short),[33] which was co-designed at MIT, is the next large space telescope aiming to discover small, Earth-like planets around the brightest stars. While its predecessor, Kepler, only monitored a small section of the night sky, TESS will survey the whole firmament. Four wide-angle cameras will continually monitor more than half a million stars and record the small variations in brightness that betray planetary transits. Kepler took 25 years to develop; TESS will find its way through the now much more exoplanet-friendly astronomical community much more quickly. Brainstorming started at MIT in 2006 and NASA plans to launch the satellite in 2017. This will be followed some seven years later by Plato, TESS's European counterpart.

Seager spends a lot of her time on the development of TESS but, at the same time, she shuttles back and forth in another direction. She picks up a kind of cake tin full of electronics from her desk. 'I had an idea when I first started at MIT,' she says. 'And it caused a kind of snowball effect. More and more people started working on it. Seven years and a few million dollars later, this is the result.' The gold-coloured object that she proudly holds in her hands is a prototype of a nanosatellite. It looks like a camera, with a lens on the short side. 'We want to launch a fleet of these,' she says. 'Each satellite will focus on one star at a time.'

The collective name for this fleet of mini-satellites is ExoplanetSat. It is a little like Kepler, with the difference that each star on the list is watched through a separate telescope. A nanosatellite is a lot cheaper to get into space than larger satellites like Kepler or TESS. And there is no need to reserve a place on an expensive rocket; nanosatellites can hitch a ride on commercial spacecraft, which are launched regularly by, for example, aircraft manufacturer Boeing. Another distinctive quality of these mini-satellites is that they are specially designed to observe the very brightest Sun-like stars. Stars like Alpha Centauri would burn through TESS's camera, but are no problem for the ExoplanetSat satellites.

IF AN Earth-like planet is indeed found around one of the bright stars, it will be relatively easy to also observe its atmosphere. 'Relatively easy' meaning at the very limits of what the next generation of spectrographs, like that on board the James Webb Space Telescope, will be able to cope with. The launch of this successor to the Hubble Space Telescope is planned for 2018, but has already been postponed several times because of the project's gigantic budget.

The second pillar of Seager's research is to observe planetary atmospheres. To do that, she has another alternative for – or rather an addition to – the James Webb Telescope. She shows me an animated film lasting about a minute, entitled *Starshade*. My jaw drops. A satellite floats through space. At one end, there is a large yellow cylinder like a kind of serviette ring around the spacecraft, the Starshade. The Starshade then disengages from the mother ship. The serviette ring rotates as about thirty flat, pointed protuberances slowly unfold in a ring, like the petals of a giant sunflower. Eventually, the flower-shaped shade flies away and comes to rest 50,000 kilometres away, in a direct line between a star and the mother ship, on which the telescope is located. In this way, the Starshade blocks the light from the star. We then see the planets that orbit the star on each side of the silhouette of the sunflower. The closer the star is, the better it works because, from our perspective, there will be more space between the planets and the star. 'If there are Earth-like planets around the ten, twenty closest stars,' says Seager, 'we will certainly find them with this.'

Starshade is a kind of space-based version of the sunshade that telescope-builder Tim Brown designed at the start of his career. The exact shape of the petals, which cover the jagged halo of the star almost completely, was designed in 2004 by astronomer Webster Cash from Colorado. The first version of Starshade was originally intended to be part of the Terrestrial Planet Finder, a combination of eight large exoplanet satellites. It was an ambitious mission – too ambitious, as NASA decided in 2011. After hundreds of millions of dollars had been spent on it, the space agency pulled the plug on the mega-project. One of the reasons was that planet hunters could not agree among themselves on the best design. Besides Starshade, a number of other instruments had been devised to obtain direct images of exoplanets.

At a symposium organized by Seager in 2011, Geoffrey Marcy openly voiced his frustration about the disagreement that had led to the Terrestrial Planet Finder project being cancelled. 'Now we have nothing,' he concluded, crestfallen. Seager herself summed it up in the words of Abraham Lincoln: 'A house divided against itself shall not stand.' Now Seager is leading a team that hopes to get Starshade into space after all, possibly as an extension to the Wide Field Infrared Space Telescope (WFIRST), set to launch in 2025.

The second film she shows me is the 'real thing'. A prototype Starshade, full size, at a NASA base in Los Angeles. Only four of the thirty petals have been fitted to the structure, but the result is already spectacular. The flower opens: the petals unfold slowly, until they extend for 30 metres. It looks like a scene from *Star Wars*: a new space weapon unfolds in a futuristic hangar with white walls. Seager beams as she looks at her laptop screen: she hasn't built a castle in the air, but a real spaceship.

THE THIRD pillar of Seager's research focuses on the last piece of the puzzle in the search for exoplanets – recognizing extra-terrestrial life. Imagine that Kepler's successor TESS or the ExoplanetSat nanosatellites find a sister planet to the Earth: a rocky planet, close enough for its atmosphere to become visible with Starshade. And imagine that we observe this planet with a spectrograph. What indicators in the planet's spectrum will betray the presence of life? Seager is trying to answer this important question by predicting biomarkers, the footprint of extraterrestrial life.

To understand how life can be detected with biomarkers, we turn the situation on its head: how would we recognize life on Earth from space? You go on board a spaceship and fly out of the solar system at an unrealistic speed. You make the same journey that we imagined Sandra Bullock taking earlier. But this time, you're not spinning around in a spacesuit, but are sitting in a comfortable capsule with a telescope on board. Through the rear window, you see the Earth and the Moon slowly getting smaller, until all that is left is a minuscule blue dot in the blackness of space. The other planets whizz past you: Mars, Jupiter, Saturn, Uranus, Neptune. Each planet that passes you is already another dot in the remote distance before the next one approaches. Lastly, you fly through the space rubble of the comet cloud that encircles the solar system for a while, until the Sun too is only a bright spot of light in the firmament.

Now you take out your telescope. The Starshade that has flown along behind you unfolds some tens of thousands of kilometres away, covering the spot of light that is the Sun. You can see the flower-shaped silhouette through the telescope, with the glow of starlight fanning out a little around it. In the faint glow, you can make out a few bright dots of light. The brightest are the two largest planets, Jupiter and Saturn, but close to the edge of the Starshade, you see four more small dots. The third dot from the Sun is the Earth. You mount a spectrograph on the telescope and record the starlight that shines through the Earth's atmosphere. In the spectrum, you see the absorption lines, the characteristic fingerprint of the molecules that occur in the Earth's atmosphere. You can easily identify water vapour, oxygen, carbon dioxide, methane and ozone.

Of the chemicals in this potpourri, oxygen, methane and ozone are examples of biomarkers, indicators of potential life. All three of these gases are produced on Earth almost exclusively by organisms, directly or indirectly. They are chemically unstable, meaning that they react quickly with various other substances. Oxygen, for example, reacts rapidly with metals in rock, causing them to rust. There would be no oxygen left at all on Earth if it were not constantly replenished.

If we zoom in with our telescope, we can see how the oxygen, ozone and methane levels in the Earth's atmosphere are maintained. Through photosynthesis, plants and micro-organisms keep the oxygen level in the atmosphere at around 20 per cent. Some of the oxygen in the atmosphere reacts again to form ozone. If the sources of oxygen on Earth are depleted, the ozone layer also rapidly disappears. The greenhouse gas methane is almost entirely produced by plants and bacteria, including the kinds of bacteria that are found in the digestive system of cows. Without bovine flatulence, there would be no methane. Methane reacts in turn with oxygen. Both would have disappeared from the atmosphere if they were not constantly replenished from the Earth. In other words, without life, the composition of the atmosphere would be completely different. With the help of biomarkers, it is therefore possible to conclude that there is life on Earth, even from a great distance.

Seager wants to apply this method to exoplanets. She has devised her own version of Drake's equation, estimating the number not of *intelligent* but of *detectable* life forms. This number will probably be greater. The SETI experiments that Drake initiated focus on signals from intelligent civilizations. Only aliens who are capable of making radio equipment can be found in this way. Biomarkers offer a greater chance of success: every life form, smart or stupid, will most probably produce certain biomarkers.

Is finding extraterrestrial life then simply a matter of identifying oxygen, methane and ozone in the atmosphere of an exoplanet? Unfortunately not. Firstly, we are not certain that these chemicals can only be produced by life on other planets, too. In 2003, methane was found in the thin atmosphere of Mars. Claims that this was proof of life were, however, quickly refuted. Methane can also be produced by chemical reactions beneath the planet's surface, without living organisms being involved at all.[34] To be certain that, when we find biomarkers, they really are a sign of life, we have to be able to exclude such processes.

Besides processes taking place on and below a planet's surface, other uncertainties make it more difficult to recognize life.

More massive planets have a different atmospheric composition – light gases like hydrogen cannot escape (like they do on Earth) because of the planet's strong gravity. Gases produced by life will react differently in a hydrogen atmosphere from how they do in a terrestrial atmosphere, eventually creating different biomarkers. And, last but not least, we have no idea whether extraterrestrial life forms produce the same unstable gases as life on Earth.

It seems like a mission impossible: to prove the existence of an unknown life form by identifying an unknown gas – and in a planetary atmosphere that is different to our own. Compare it to what happens when you receive a text message on your mobile phone. You know a message has arrived because you have selected a specific ringtone for receiving messages. But what happens if the same ringtone goes off on your brother's phone? He may have set the same tone for his email. So the same signal can be caused by different things. This 'ringtone principle' also applies to observing biomarkers in the atmosphere of an exoplanet. The presence of methane in a planet's atmosphere does not necessarily mean that space cows are passing wind down on its surface.

We came across a similar way of thinking earlier in Christiaan Huygens's book *Cosmotheoros*. Huygens noted that the ingredients for life on Earth did not necessarily provide a blueprint for life on other planets, saying 'Perhaps their Plants and Animals may have another sort of Nourishment there.' Three centuries later, Seager is one of the scientists who dares to take up Huygens's challenge. She poses the same question in a new form: 'I have no idea what extraterrestrial life will look like; endless guessing gets you nowhere at all. So instead, we pose the question from the perspective of the observer, asking what biomarkers we can detect at all.'

In the search for new biomarkers, Seager pretty much empties the whole chemical Lego box out on the floor. First, she looks at the most common atoms in organic molecules: carbon, hydrogen, oxygen, nitrogen, phosphorous and sulphur – the 'big six', as she calls them. She considers all possible molecules that you can make from these six basic elements, up to a specific number of atoms per molecule – otherwise it is too complex for today's computers. Then she passes all the possible permutations through a sort of chemical filter, so that only those gases are left over that are eligible as biomarkers: unstable and not produced by geophysical processes. Three hundred years after Huygens, we are finally working on how we can prove the existence of extraterrestrial life.

SEAGER DOES not have to achieve her ambitious research agenda on her own. Her group at MIT consists of some fifteen students and researchers. Not all of them have a background in astronomy – there are also, for example, aeronautical engineers and geophysicists. I ask her how she selects her students, how she recognizes talent. 'I was talking to a student who was here for a summer project,' she tells me, by way of example. 'I tossed a few of the big questions that I was working on at him. Most students aren't interested in projects without clearly defined limits, it's too much for them. But with him, I saw something . . . the twinkle in his eyes, the way he spoke about the future. John Bahcall could pick up on that every time; I learned it from him.

'Here at MIT, you find the world's best. I don't mean that arrogantly, it's just the mentality here. I don't need to teach my students physics. But other things are more difficult for them. How do you choose a good problem? How do you solve it? How do you deal with politicians, market your research, think big? Those are the kind of things I try to teach them.

'Afterwards, everyone does things differently than I had expected. When my two sons play with Lego, one sorts all the bricks neatly and builds a tower of each colour. The other one builds a chaotic structure that goes off in all directions, but is still stable. Should I then teach one to be a little more creative and the other to work in a more structured way? Neither, I think. And the same applies to my students. I help them to expand on the talents they were born with. Some people discover them themselves and others need a little push.'

Seager talks about her students with the same affection as for her children. After completing a Masters or a PhD, many of them receive grants to continue their research at another prestigious university. 'One of my first doctoral students, Nikku Madhusudhan, now lectures at Cambridge. Yeah, Cambridge,' she repeats with undisguised pride.

IN A nearby office, Seager introduces me to her research group. Andras and Vlada, two young researchers, shake my hand. We talk to each other for an hour; the door opens continuously and other members of the group, Mary, Stephen and Julien, introduce themselves. I am charmed by the optimistic American mentality of these young people. Europe and the u.s. differ widely in their approach to science and especially in how it is presented to the outside world. Imagine a young European researcher bent over a large table where, in the light of a desk lamp, he is very carefully gluing two matches together. When asked what he is doing, he will probably explain in detail that, although the two matches are securely joined together, there isn't a drop of glue to be seen. He will tell us that he is using a mixture of two kinds of glue for the best result. Then he uses a precision spray to apply the glue. He shows how he repeats this process for each match. Only when asked what it is, will he say that the pieces of glued-together wood will eventually be a miniature warship. It will then be fitted with sails, just like a real ship. Maybe he will even get it to float – at least if he can find a good impregnating agent.

His American counterpart is more likely to open the door himself and invite you to come and see what he is making. He is building a ship! The ultimate aim is for it to be seaworthy, and it is to be fitted with miniature sails. He tells you where it is going to sail to and how it can be upgraded; he might even fit it with wings. If you look over his shoulder, all you see are the same two matches you saw on the European's table.

The young researchers in Seager's group have clearly adopted the American way of presenting their work. They start with the desired outcome of their research and only then fill in the details. They don't mention a lot of ifs and buts. They all do something a little different, but they clearly have one shared goal – to find extraterrestrial life.

Mary is working on the prototype of the ExoplanetSat nanosatellites that are designed to find new planets. Julien has developed a new method for observing the spectrum of a planet's atmosphere. Stephen is exploring what potential biomarkers will be the easiest to observe in that spectrum. Andras and Vlada are investigating what phenomena, besides life, also leave traces in a planet's atmosphere, such as continental shift, the climate and cloud formation. The group conducts experiments in a cloud chamber, a laboratory a little like a smokers' bar, where clouds are created using all kinds of chemicals that may be found on other planets.

I am dazed by all the information. At the start of the day, I was already impressed that there are astronomers who study the atmospheres of exoplanets. Now I'm sitting in a room with people who simulate extraterrestrial volcanoes, oceans and tornados on their computers and in the lab. Scientists overflowing with energy, dynamism and self-confidence.

THE GROUP at MIT is of course not the only one in the world that is using innovative methods to tackle the big question. Astrobiology is a growing specialism and new research groups are cropping up everywhere. Five kilometres away at Harvard,

Seager's former boss Dimitar Sasselov is leading the Origins of Life initiative, an interdisciplinary team of astronomers, geophysicists and biochemists. One of their experiments is to create a living cell from separate organic components. In 2014, another group, at University College London, succeeded in simulating methane in hot planetary atmospheres, so that observers know where in the spectrum they need to look to detect this biomarker.

In space, too, there is a lot on the planning board. Tim Lee, chief of NASA's Space Science and Astrobiology Division, told me in a cafeteria in Mountain View about the many space missions in preparation to search for exoplanets and signs of life. And the Netherlands is also playing its part. Leiden professor Ignas Snellen has developed an original and inventive method of observing planetary atmospheres from the ground, which is cheaper and less risky than using space telescopes, without having to wait for the planet to pass in front of its star. Snellen – who, according to his website, would rather have been a professional footballer than an astronomer – used the same method in 2014 to be the first to measure the rotation of an exoplanet on its own axis.

After spending an afternoon with Sara Seager and her group, however, I have been ignited by their enthusiasm. It is difficult to believe that they will not achieve their goal, despite the fierce competition. I would not be at all surprised if the next breakthrough in the search for extraterrestrial life is made by the space rebels on the seventeenth floor of the light-grey tower overlooking the Charles. On their own, using their own, sometimes quirky, methods, from within the rebellious MIT.

eleven

THE PLANETARY CIRCUS

THE SUN is shining on the Stadthalle in Heidelberg, Germany, the conference centre on the River Neckar. For a whole week, in July 2013, the red brick building is the venue for the sixth 'Protostars and Planets' conference, the place to be for astronomers studying the origins of stars and planets. Almost a thousand of them come from all over the world to present their research. Specialists talk about the countless important developments and discoveries in the discipline since the previous conference six years earlier. Later, all their presentations will be published in a thick book, which will serve as the Bible for those working in the field for the coming six years.

Even more interesting than the lectures in the main hall are the research posters, hundreds of which hang everywhere in the large complex. Almost every researcher, young or old, has brought one along. They show what is happening at that moment on the front line. With a cup of coffee in one hand and gesticulating wildly with the other, astronomers tell each other about their latest discoveries, results that are so fresh that they haven't made it to the scientific journals yet. It is sometimes sensitive information, because it is a competitive field. The posters don't reveal all the data, but they tell you what conclusions the authors have drawn from them.

One room is permanently full to bursting: the planet room. Men and women wriggle through the crowd in all directions, check out the posters and sometimes chat to the authors. I try to find Lisa Kaltenegger, a German planet hunter who leads a group in Heidelberg and is a visiting professor at Harvard. Her poster shows eight Earth-like planets, which she wants to observe in greater detail with a new generation of telescopes. But it proves impossible to speak to her for longer than a minute, as rivals or friends buzz around her.

A Princeton student stands proudly in front of her own poster, about exomoons – moons around exoplanets. That is to say, a moon that orbits a planet that orbits a distant star. None have yet been discovered but no one doubts their existence; after all, most planets in our solar system have moons. She has an enthusiastic answer ready when I ask her why we should search for exomoons: 'If we have learned one thing, it is that we should never rely on our intuition when it comes to new discoveries. We don't know what the search for new moons will come up with. It might be much better to live on a moon than on a planet.'

That may very well be true. The moons in the solar system are also rocky. None of them, except Titan, Saturn's largest moon, has any atmosphere to speak of, but their composition and surface temperature are comparable to those of rocky planets. And even on Earth, we know of places where life can exist under such extreme conditions. Lake Vostok, for example, is a subterranean lake near the South Pole. It has been covered by a layer of ice kilometres thick for millions of years. On the surface, the temperatures are barbaric, up to 90 degrees below zero. But the underside of the ice layer has melted, probably due to heat from volcanic material beneath the Earth's crust.

In 2012 a Russian drilling team started bringing up small quantities of water from the lake. A year later, it was clear that Lake Vostok was teeming with life. More than three thousand different micro-organisms were found in the water samples: not only a wide variety of bacteria but multi-cell organisms related to molluscs and fish. The results showed that life on Earth can survive in extreme circumstances.

The conditions in Lake Vostok were compared with those on Europa, one of Jupiter's moons. Europa, which was discovered by Galileo, is about four times smaller in diameter than the Earth and is completely covered in ice. Scientists suspect that, below this outer layer, there is a large ocean. In 2013 and again in 2016, the Hubble Space Telescope discovered spectacular geysers on Europa's surface that spray ice particles and water vapour hundreds of kilometres into space.

The same applies to Saturn's moon Enceladus, which is much smaller than Europa (its surface area is a little bigger than Turkey) and is also covered in ice. The spacecraft Cassini photographed geysers on Enceladus in 2005. This makes the two moons interesting destinations for future space missions, which may be able to find out whether – like in Lake Vostok – there are living organisms in their cold, subterranean oceans.

A robot mission like the Curiosity Mars Rover, which could actually land on the surface of the moons, only exists so far in the PowerPoint presentations of optimistic space technicians. But the European Space Agency (ESA) has plans to launch the Jupiter Icy Moon Explorer (JUICE for short; a new high point in the history of funny names for spacecraft and astronomical instruments). Over a period of a few years, JUICE will perform close flybys of three of Jupiter's moons, including Europa. The mission's aim is to learn more about the conditions under which life can evolve and survive.

JUST AS at MIT, I am impressed at the optimism and energy buzzing through the room in Heidelberg. It is like one big planetary circus. I see posters with exotic titles like 'The Habitable Zone around Binary Stars' and 'The Structure of Ice Planets'. Doctoral students, researchers and professors chat animatedly about a proposed observation campaign, a new article or good schnitzel restaurants in Heidelberg.

This is what a new discipline looks like – a room full of astronomers in shorts, T-shirts and sandals engaged in lively conversation about their shared passion. Everyone wants to be a part of it. 'Exoplanetary science is a party you really want to gatecrash,' a colleague from a different field said to me. New telescopes have reached the Goldilocks zone. Rocky planets are in orbit around stars that give them similar temperatures to those on Earth. Gases have been discovered in their atmospheres, and new telescopes may even detect biomarkers. The search for extraterrestrial life is approaching its climax.

MANUEL GÜDEL, an Austrian professor with a grey moustache and a characteristic accent, is the final speaker at the conference. He brings us all firmly back to Earth. In an hour-long talk, he outlines the conditions that make life on our planet possible. Looking at each one separately, he compares these conditions to those on other stars and their planets and shows how favourable these places are for the emergence of extraterrestrial life. He emphasizes that he is not going to try and tell us what extraterrestrial life might look like. 'Then I would have to fantasize about beings with big eyes and green teeth. Nor am I talking about how life originates. I am purely concerned with the necessary conditions, the starting conditions that determine the habitability of a planet.'

A nocturnal stargazer might imagine, while looking at the vastness of the universe, that it must be teeming with life. However, Seager's Equation, which estimates the number of planets on which life occurs, still contains too many uncertainties. We do know more about some of the factors Frank Drake included in his original formula in 1961, but they mainly relate to the number of planets in the Milky Way. The Kepler mission has now shown us that each star has an average of one planet. This may even be a conservative estimate, because smaller planets are more numerous. Perhaps we will also know in the not too distant future how many of these planets occur on average in the habitable zone of their stars. If the trends we have detected so far also prove to apply to the Goldilocks zone, this number can be expected to be substantial.

But that is where the great uncertainty starts, Güdel tells us. It is very difficult to estimate the percentage of habitable planets in the Goldilocks zone that are also actually inhabited. The temperature on the surface of a planet is not only determined by its distance from a star. Greenhouse gases can cause it to rise quickly and create uninhabitable conditions, like those on Venus.

Güdel makes a sobering summary of other important factors for habitability. Firstly, there is the stability of planetary orbits. In our solar system, the gravitational pull of Jupiter, the most massive planet in the solar system, acts as a kind of anchor on all the other planets. Without a massive planet, the orbits of the smaller planets would be much more chaotic and unstable. That would probably cause them to have irregular climate patterns and seasons, providing a much less comfortable environment for life. No Jupiter, no life, it would seem.

Furthermore, the influence of the star itself should not be ignored. A star is not a lamp that continues to burn with the same power for billions of years. Since it was born, the Sun has gradually emitted about a third more light, which has significantly affected the surface temperature of the planets, and therefore their habitability. A rise in temperature on Earth of only a few degrees has a considerable impact on the climate and life on the planet, including the extinction or emergence of whole species.

Güdel, who specializes in ultraviolet and X-ray radiation, goes into detail about the harmful radiation from the Sun, which is kept at bay by the ozone layer around the Earth. We know very little about how this works in the case of other stars. Three-quarters of stars are red dwarfs, cool stars that emit relatively high levels of X-ray radiation. Their low temperatures mean that their Goldilocks zone is relatively close and planets in the zone will probably be more exposed to radiation. Second, because of the strong gravity at such short distances, these planets always show the same face to their stars (like the Moon to the Earth). Consequently, it is constantly day on one side of the planet and night on the other and this, too, considerably affects their habitability. Nevertheless, it would certainly be wonderful to see the eternal sunset in the twilight zone between the two hemispheres.

Güdel continues. He talks about the solar wind, a flow of particles that is kept away from the Earth by the planet's magnetic field. The solar wind and the magnetic field strongly influence the composition of the Earth's atmosphere, but we know very little about how both phenomena behave in other planetary systems.

Then, of course, there is the Moon. Our Moon was likely created following a collision between the Earth and a large asteroid. Since then, it has been responsible for the tides, and ensures that the Earth's axis does not wobble but remains stable at the same angle, giving us a regular cycle of seasons. Without that random collision, many of the conditions that make the Earth habitable would probably have been very different. If there are still so many great mysteries in our own solar system – was there ever life on Mars? why does Venus have so many greenhouse gases? – how can we ever predict the habitability of exoplanets?

After listening to Güdel's list of conditions for life to emerge, my optimism is somewhat tempered. If you look back at the history of the Earth, a lot of switches just happened to be set in the right position to make the origin and evolution of life possible. That doesn't necessarily mean that, if one of those switches had been in the wrong position, life would have been impossible. Perhaps, via a random walk through the forest of possibilities, a liveable situation would somehow have been created. There are two possibilities – either life is everywhere, or Earth is a one-off. But the motto stays the same: keep asking questions, keep hunting.

IN MY hotel room that evening I watch a report on the conference on the local news. A German professor is talking about the latest planetary discoveries and gives the detection of water vapour in the atmosphere of an exoplanet as an example. Outside, they ask the man in the street for his opinion. The interviewer stops a small man with a cap as he is walking past the Stadthalle. He thinks it is ridiculous. 'Why are hundreds of people in there getting so worked up about a bit of water vapour?' he asks incredulously. 'No one cares about that, do they? They should be out here in this weather!'

Water vapour was, of course, not the only thing that the astronomers discussed in the sun-baked Stadthalle, but the man in the cap had a point. Why do we do this, actually? Why are we so fascinated by the existence of extraterrestrial life? Why do we want to know whether, on some far-off planet that we can never hope to reach, there is something that moves, breathes and reproduces itself? Dutch comedian Theo Maassen describes it very aptly:

> Take space travel. Billions are invested in it. Billions! And, I always think, what for? Why? Okay, they want to know if extraterrestrial intelligence exists. But then I think, there aren't that many options. It either does or it doesn't. That's it. If it doesn't, I think it's a waste of all that money. If it does, there are two possibilities: either they're dumber than we are, or they're smarter than we are. If they're dumber, I don't want anything to do with them. And if they're smarter, they'll find us before we find them!

I have also sometimes asked myself what is so fascinating about the search for extraterrestrial life, intelligent or not. In the twentieth century, astronomers went looking for the most extreme parts of the universe, where our knowledge of the laws of nature was put severely to the test. We became enthralled by the enormous density in the interior of neutron stars, the extreme gravity close to black holes and, of course, the Big Bang, the origin of the universe and everything within it. So what's so interesting about looking for a small chunk of rock in orbit around an average star, just because there might be complex molecules on it somewhere? It doesn't tell us anything new about the laws of nature, the origins of the universe, the great mysteries of existence, does it? Why look for a needle in a haystack when you can spend your time learning more about the haystack itself?

The planet hunters themselves have an answer to that. David Charbonneau, the Harvard professor with the nose for a scoop, thinks what makes hunting planets so special is that so many of their fundamental properties can be directly derived from the observation data. A Doppler measurement gives the mass of a planet, a transit measurement its size. We don't have to make many assumptions – it's all pure science.

Bill Borucki, the tenacious inventor, puts his life's work in perspective:

> In thirty years' time, everyone will have long forgotten that there was a satellite called Kepler. They'll only remember what the mission discovered: that there are planets around every star. That alone will ensure that we get just enough support for high-risk projects to get us over the next cliff and discover new things.

Didier Queloz, co-discoverer of the first exoplanet, describes the feeling that finding a new planet gives you. 'I call it *planet fever*,' he says. 'It does something to you. You defend it as if it were your own child. It's like you've planted a flag on a new world.'

Sara Seager often asks herself why people are so keen to believe in extraterrestrial life. She also worries about whether biomarkers will give us a conclusive answer. 'Will people ever be satisfied that a biomarker gas on a small dot somewhere very far away is proof of life, rather than a visit from a real alien?' she wonders. She thinks that, if we do find an exoplanet with the right biomarkers, it will be a good idea to immediately point a radio telescope at it and listen for possible signals from an intelligent civilization. We may even be able to send a message back. An idea that Jill Tarter and Seth Shostak would certainly endorse.

Planet fever had scientists in its grasp long before the first exoplanet was found, and long before the start of the century that Giordano Bruno had predicted. From Tycho Brahe's precise measurements of the orbit of Mars to the model that Kepler derived from them, from Christiaan Huygens's speculations about aliens to the failed planetary discovery of Piet van de Kamp – they were all infected with the bug. In his small office in Berkeley, Geoffrey Marcy, the larger-than-life performer who is not one to shy away from big, lyrical statements, describes the motives of the planet hunters in glowing terms.

> We scientists often pretend to be objective, that we only study the really important questions. We pretend that we switch off our emotions, and try to solve fundamental problems as logically as possible. But that's not true. No one can say objectively what questions are the most important to ask. We come up with questions that we like, in the same way that we like one of Beethoven's symphonies. The way we devise scientific questions cannot be defended rationally, they form like poetry in our minds. We love planets, so we study them. We think it's important, but we can't prove it. It is important to *us*. There's nothing wrong with that. In fact, we should cherish it, that's how humanity has always been. Curiosity has helped our species to survive, and curiosity will always help us to move forward.

Marcy's words remind me of Carl Sagan, the legendary spokesman for astronomy who brought the search for extra-terrestrial life to the attention of the public at large. In 1990, thirteen years after its launch, the spacecraft *Voyager 1* had just passed Pluto and was about to leave the solar system. Sagan suggested getting the little probe to look back just one more time, to where it had come from. From a distance of six billion kilometres, Voyager took a photograph of the Earth. You have to do your best to see it but, in the black, grainy image, a small blue dot is just visible. Sagan's reflections on this 'pale blue dot' became famous as an exercise in perspective:

> From this distant vantage point, the Earth might not seem of any particular interest. But for us, it's different. Consider again that dot. That's here. That's home. That's us. On it everyone you love, everyone you know, everyone you ever heard of, every human being who ever was, lived out their lives . . . It has been said that astronomy is a humbling and character-building experi-ence. There is perhaps no better demonstration of the folly of human conceits than this distant image of our tiny world.

The hunt for planets and extraterrestrial life cannot be justified objectively. We are fascinated by other life forms, we want to compare them with ourselves. We want to look further, searching for another small dot somewhere far away, around another sun. We want to know who lives there, and how they live. Think about those old maps of the world from a time when large parts of the globe had not yet been discovered. On the edges of the map, in oceans that had not yet been charted, there were drawings of sea monsters and the text *Hic sunt dracones*, 'Here be dragons'. The unknown fills us with fear, but also has an irresistible attraction. We hunt for new planets for exactly the same reason that ancient explorers braved the sea dragons. Our sight may be defective, but our minds are incurably inquisitive.

It may happen in this century: the discovery of life elsewhere in the universe. Perhaps there will come a day when I can take my own grandchild to Teylers Museum. By then, the oval room will be full of orreries. Thousands of copper balls with marbles of all shapes and sizes circling them. On that day, I will point at one of those marbles and tell my grandchild that there is life on that planet. And then the questions will come, thick and fast.

GLOSSARY OF TERMS

Absorption line – Specific wavelength in the spectrum where starlight is absorbed by atoms and molecules in the outer layers of the atmosphere of a star or its planets. Absorption lines thus create a 'fingerprint' of the chemicals present in the atmosphere.

Asteroid – A chunk of rock in orbit around the Sun, smaller than a comet.

Astrobiology – Also known as 'bioastronomy'. The study of the origin and existence of life in the universe.

Astrometry – Branch of astronomy concerned with the precise measurement of the positions of celestial objects (stars, planets, etc.).

Astronomy – Study of extraterrestrial natural phenomena. Not to be confused with astrology (drawing up horoscopes).

Biomarker – Unmistakable sign of life in a planet's atmosphere; the presence of a gas that can only be produced by life.

Brown dwarf – 'Almost-star': globe of gas not quite hot enough for nuclear fusion to occur in its interior. Brown dwarfs are considerably larger than planets – ten to a hundred times the mass of Jupiter.

Comet – Rocky object usually tens of kilometres across in a wide and elliptic orbit around the Sun and which occasionally crosses the orbits of the planets. Comets have a characteristic 'tail', created when solar radiation causes gas and dust to stream away from the comet's surface.

Doppler effect – Or Doppler shift. The phenomenon by which light waves are compressed or elongated as their sources move towards or away from the observer. A light source that is moving away appears redder than when it is immobile, while a source that is approaching appears bluer. The Doppler effect also causes the absorption lines in the spectrum to shift.

Doppler method – Method that uses the Doppler effect to detect the presence of exoplanets. A 'wobble' in the motion of a star caused by the presence of a planet is measured by observing the Doppler shift of its absorption lines. See the illustration on page 125.

Dwarf planet – Small planet-like body that has not 'swept' its orbit clean, so that multiple similar objects are found in the same orbit.

Earth-like planet – Planet comparable in size and mass to the Earth.

Exoplanet – Planet around a star other than the Sun.

Galaxy – A large cluster of hundreds of billions of stars, held together by gravity. Most of the stars in the universe are found in galaxies. The galaxy in which our own solar system is located is known as the Milky Way.

Hot Jupiter – Gas planet close to a star, so that its surface temperature is very high.

Light year – The distance travelled by light in a year: 9.46 trillion (9,460,000,000,000) kilometres.

Meteorite – A small chunk of rock from space that impacts on the Earth. A meteorite that has not yet impacted is known as a meteoroid.

Molecule – Chemical compound of two or more atoms.

Planet – Near-spherical celestial body of rock and/or gas, in orbit around a star (e.g. the Sun). A planet is the largest body in its own orbit; multiple small objects in the same orbit are known as dwarf planets.

Planetary system – Collection of planets around a star.

Red dwarf – Star with a mass of 10 to 50 per cent of that of the Sun and a surface temperature of 2,000–3,000 degrees Celsius. Three-quarters of all stars in the Milky Way are red dwarfs.

SETI – Search for Extraterrestrial Intelligence: movement that searches for radio signals from extraterrestrial civilizations.

Solar system – Collective name for the Sun and its planetary system.

Spectral line – See absorption line.

Spectrograph – Instrument that breaks down the light beam entering a telescope, so that the colours of which the light is comprised can be detected.

Spectrum – The colours of which light is comprised. The spectrum can be made visible by breaking the light down. A rainbow (which is created by sunlight broken down by raindrops) is a good example.

Star – Sphere of hot gas, held together by gravity, and in the centre of which hydrogen is converted into helium by nuclear fusion. The energy released during this process is emitted as (light) radiation.

Sun – Star at the centre of the solar system, with a surface temperature of 5,500 degrees Celsius.

Super-Earth – Exoplanet larger and more massive than the Earth (up to approximately ten times the mass of the Earth).

Telescope – Optical instrument that makes distant objects visible using lenses or concave mirrors. Instruments that observe other forms of radiation than visible light (e.g. gamma rays, radio) and do not use optical lenses are also referred to as 'telescopes'.

Transit method – Method of detecting the presence of an exoplanet by recording a temporary reduction in the brightness of a star caused by a transit (the planet moving in front of the star). See the illustration on page 124.

Universe – All of space and everything within it, including stars, planets, galaxies and the interstellar medium.

REFERENCES

1   In 2006, the International Astronomical Union (IAU), which
    decides on important questions of this nature, definitively
    demoted Pluto to the status of a dwarf planet. Since Pluto had
    been discovered, many more similar dwarf planets had been
    found in the same region of the solar system and the IAU didn't
    want to classify all of them as planets willy-nilly. The decision
    caused a lot of controversy and indignation: many astronomers
    were very attached to the tiny lump of rock. A professor at my
    institute still has a bumper sticker on the door of her room
    saying 'Honk if you think Pluto is still a planet.' All objections
    notwithstanding, the solar system currently has eight full-fledged
    planets.

2   In 2013 Canadian punk band Crusades made an album in honour
    of Bruno, with one rendering of his famous statement as its title:
    *Perhaps You Deliver this Judgment with Greater Fear than I Receive
    It.* Ear plugs are recommended.

3   See the links on www.planetenjagers.nl for an animated image
    of the apparent movement of Mars.

4   In Luther's time, astrology (making horoscopes) and astronomy
    had not yet become separate disciplines.

5   Elpino means the six planets observable with the naked eye, plus
    the Moon.

6   Newton only published his laws of gravity when urged to do so
    by astronomer Edmund Halley.

7   The ring of gas around Saturn had been observed earlier in the
    seventeenth century by Galileo but, because of the bad quality
    of the image through his telescope, he thought he was seeing
    moons on opposite sides of the planet.

8   Richer's estimate was 10 million kilometres less than the
    distance established using modern methods. That sounds like

a serious miscalculation but, relatively speaking, he was only out by less than 10 per cent. If you work with astronomical numbers regularly, you don't take much notice of a million here or there.

9   For the sake of convenience, we measure the scaled-down distance to Alpha Centauri over the curved surface of the Earth.

10 At this moment, Proxima Centauri is actually the closest star to the Sun. It will take another 25,000 years for it to pass behind Alpha Centauri A and B, and they will once again become the closest stars.

11 The brightest stars are named after the constellation in which they are found. They are first ranked according to brightness using Greek letters: Alpha Centauri, for example, is the brightest star in the constellation Centaurus, and Beta Pictoris the second brightest in the constellation Pictor. When the Greek letters are used up, numbers are used, but to add to the confusion, these numbers refer to their relative positions instead of their brightness. Fainter stars mostly have a catalogue name. One of the most well-known and widely used catalogues is the Henry Draper catalogue (named after an American amateur astronomer who had little to do with compiling the actual catalogue, which was done 25 years after his death by the Scottish-born maid-turned-astronomer Williamina Fleming) are designated with the prefix HD and a number with six or seven figures, for example, HD 163296.

12 This derisive name for extraterrestrial beings (where the derision is aimed at those who are looking for them) was used as long ago as the nineteenth century for extraterrestrial beings in fairy tales.

13 Planetary systems have since been found around both Epsilon Eridani and Tau Ceti; astronomers continue to seek signs of life around the two stars.

14 The actual estimate was not the total number of stars in the Milky Way, but the number formed per year. That takes account of the fact that not all stars are the same age.

15 There may be several reasons for the slow rotation of the Sun and other stars. Besides the transfer of the angular momentum to planets, stars are for example also slowed down by the magnetic field of their surrounding birth cloud. Either way, Struve was right in assuming that most star systems have planetary systems.

16 The illustrations on pages 124 and 125 show these two methods. Struve's visionary publication was actually preceded by French astronomer David Belorizky, who in 1938 similarly suggested both the Doppler and transit methods for the detection of exoplanets. His publication being written in French might have been a reason for it having been omitted from canonical exoplanet histories.

17 In his book *Planetenbiljart* (Planet Billiards), Gerard 't Hooft recalls an anecdote about Sagan that illustrates his impact on science. One evening in 1965, a few years before the moon landing, Sagan was in a bar near Harvard with his colleague Sidney Coleman, enjoying his third whiskey. They were wondering whether the astronauts that would be coming back from the Moon in the near future would bring any potentially harmful bacteria with them. They thought it would be a good idea to put Moon returnees into quarantine for a week. Coleman had long forgotten that evening's conversation when, some time afterwards, he received a draft article from Sagan by mail entitled 'Spacecraft Sterilization Standards'. When NASA later started to think about the same question, this article was the only one they could find in the literature. Sagan and Coleman's guidelines were adopted and a special Lunar Receiving Laboratory was even set up where returning astronauts looking forward only to a cheese sandwich and a night in their own beds were shut up for a week in a sterile area. 'And so the fate of mankind was decided after a few glasses of whiskey in a bar in Cambridge, Massachusetts,' Coleman concluded.

18 A CCD (Charge Coupled Device) uses a small electrical charge to detect light. It was invented in 1969 by George E. Smith and Willard Boyle, for which they were awarded the Nobel Prize for Physics forty years later, in 2009.

19 Galileo first observed the small moons in Padua in January 1610. He moved to Florence in the autumn of the same year.

20 Francesco Palla passed away unexpectedly in January 2016, while on his way to an astronomy conference.

21 The International Astronomical Union, the body that officially assigns the names of planets and is infamous for its unpopular decision to deprive Pluto of its planetary status, obstructed the giving of informal names to exoplanets for many years. In August

2013 this ban was relaxed and now anyone can submit a proposal to name a planet. There are still restrictions, however: names with political or military connotations are forbidden, as are the names of pets. After a contest to name 51 Pegasi b, it is now known officially as 'Dimidium', the Latin word for 'half'. This is a reference to the planet's mass, which is half that of Jupiter. It is arguably a less imaginative name than the runner-up in the contest, Bellerophon, after the Greek hero who tamed the flying horse Pegasus.

22 In 2015 Marcy retired from the astronomy faculty at Berkeley, after its Title ix office found that he had violated the university's sexual harassment policy.

23 To temper expectations, some astronomers prefer to use 'temperate' rather than 'habitable' zone. Without observing the planet's atmosphere, there is no way of knowing whether liquid water – let alone life – could actually exist on the surface.

24 By the time this book went to print, these numbers were likely already out of date. This shows how active a field planet hunting is. The current number of confirmed exoplanets can be found on exoplanets.org.

25 You can see the film of the Kepler Orrery by clicking on the link at www.planethunters.org.

26 The indefinite article 'a' cannot be heard on the transmission. But without it, the sentence is not entirely logical, as 'man' and 'mankind' mean the same thing. Many people, however, think that it sounds more poetic without the 'a'. Armstrong himself always claimed that he said 'a man' and that an error in the transmission made it inaudible. Recent audiographic studies and analyses of Armstrong's body language and accent (he came from Ohio) have not proved conclusive one way or another. For the time being, it is seen as a poetic slip of the tongue that most people are more than prepared to forgive the first man on the Moon.

27 Infrared detectors have to be kept very cold, hundreds of degrees below zero. This is because, at room temperature, a detector emits infrared radiation and will therefore measure itself rather than the space clouds it is directed at.

28 Besides hydrogen, carbon monoxide (co, one carbon atom and one oxygen atom), ammonia (NH_3, one nitrogen, three hydrogen)

and water (H_2O, one oxygen, two hydrogen) were found in star formation clouds in the 1960s.

29 A terrestrial example of such a photochemical process is the formation of vitamin D, a molecule in the body that is activated by exposure to sunlight. Photosynthesis, the production of oxygen by plants, is also set in motion by radiation from the Sun.

30 This impressive drawing can be seen at http://water.usgs.gov/edu/earthhowmuch.html and on www.planetenjagers.nl.

31 I have also experienced this frustration myself. I have been to the Atacama Desert four times to conduct observations for my own research. On two of these trips, the hailstones clattered down on the roof. Instead of using my costly observation time to gaze at the night sky, I was forced to spend the nights drinking coffee and enjoying Chilean scientists' particular brand of humour.

32 Hubble's most important unexpected discovery, for which the research team was awarded the Nobel Prize in 2011, is that the universe is expanding increasingly rapidly. Thousands of books can be (and have been) written on this discovery; here we restrict ourselves to the stories of the planet hunters.

33 Despite my unremitting fascination for the origins of astronomical acronyms, I have been unable to find out which Tess this satellite is named for. I'm assuming it is the daughter or mother-in-law of one of the designers.

34 An example is serpentinization, a reaction between water and rock.

BIBLIOGRAPHY

Feynman, Richard P., *The Character of Physical Law* [1965] (Boston, MA, 1994)

Introduction: New Marbles

Carlin, George, *Napalm and Silly Putty* (New York, 2001)
Lévi-Strauss, Claude, *Le Cru et le cuit* (Paris, 1964)

1 The Century of Bruno

Brahe, Tycho, *De Nova Stella* (Copenhagen, 1573)
Bruno, Giordano, *De l'infinito, universo e mundi* (London, 1584)
Bryson, Bill, *A Short History of Nearly Everything* (London, 2003)
Copernicus, Nicolaus, *De Revolutionibus orbium coelestium* (Nuremberg, 1543)
Drake, S., 'Copernicanism in Bruno, Kepler and Galileo', in *Copernicus: Yesterday and Today*, ed. A. Beer, K. Strand (Oxford and New York, 1975)
——, 'Proceedings of the Commemorative Conference in Washington DC in Honour of Nicolaus Copernicus', *Vistas in Astronomy*, XVII (1975), pp. 117–90
Dreyer, J.L.E., *Tycho Brahe* (Edinburgh, 1890)
Gingerich, Owen, 'The Great Martian Catastrophe and How Kepler Fixed It', *Physics Today* (September 2011)
——, *The Book Nobody Read: Chasing the Revolutions of Nicolaus Copernicus* (New York, 2004)
Koestler, Arthur, *The Sleepwalkers: A History of Man's Changing Vision of the Universe* (London, 1959)
Kuhn, Thomas, *The Copernican Revolution* (Cambridge, 1957)

Lederman, L.M., *Chicago Tribune* (20 October 1988)

Lockwood, Thomas P., 'How Fact Becomes (Anti-Catholic) Fiction', *Catholic Answers Magazine*, xx/8 (2009)

McMullin, Ernan, 'Giordano Bruno at Oxford', *Isis*, LXXVII/1 (1986)

Rosen, Edward, *Copernicus and his Successors* (London, 1995)

Rowland, Ingrid D., *Giordano Bruno: Philosopher/Heretic* (Chicago, IL, 2009)

2 The Little Sand Reckoner

Andriesse, C. D., *Titan kan niet slapen: een biografie van Christiaan Huygens* (Amsterdam, 1993)

Christianidis, J., D. Dialetis and K. Gavroglu, 'Having a Knack for the Non-intuitive: Aristarchus's Heliocentrism through Archimedes's Geocentrism', *History of Science*, XL (2002), pp. 147–68

Huygens, Christiaan, *Kosmotheoros, sine De Terris Coelestibus, earumque ornato* (The Hague, 1698)

Huygens, Constantijn, and Theodore Jorissen, *Mémoires de Constantin Huygens* (The Hague, 1873)

——, *Oeuvres Complètes*, vol. I, no. 5

Olmsted, John W., 'The Voyage of Jean Richer to Acadia in 1670: A Study in the Relations of Science and Navigation under Colbert', *Proceedings of the American Philosophical Society*, CIV/6 (1960), pp. 612–34

——, 'The Scientific Expedition of Jean Richer to Cayenne (1672–1673)', *Isis*, XXXIV/94 (1942), part 2, pp. 117–28

Galileo Project, http://galileo.rice.edu/Catalog/NewFiles/richer.html

Romein, Jan, and Annie Romein-Verschoor, 'Christiaen Huygens: Ontdekker der waarschijnlijkheid', in *Erflaters van onze Beschaving* (Amsterdam, 1977)

Touchard-Lafosse, Georges, *Chroniques de l'oeil-de-boeuf* (Paris, 1864), vol. III, no. 22

Voltaire, *Zadig ou la Destinée*, first published as *Memnon* (Amsterdam, 1747)

3 An Inquisitive Mind and Defective Sight

Beets, Nicolaas, *Mannekens in de Maan*, 1836, from *Dichtwerken van Nicolaas Beets* (Amsterdam, 1876), part 1, pp. 393–402

Fontenelle, Bernard Le Bovier de, *Entretiens sur la Pluralité des Mondes* (Paris, 1686)

Frederick, L. W., 'Peter van de Kamp (1901–1995)', *Publications of the Astronomical Society of the Pacific*, CXIII (1996), p. 556

Hirshfeld, Alan W., *Parallax: The Race to Measure the Cosmos* (New York, 2002)

Jacob, W. S., 'On Certain Anomalies Presented by the Binary Star 70 Ophiuchi', *Monthly Notices of the Royal Astronomical Society*, xv (1855), p. 228

Kamp, P. van de, 'Astrometric Study of Barnard's Star', *Astronomical Journal*, LXVIII (1963), p. 515

Kawaler, A., and J. Veverka, 'The Habitable Sun: One of William Herschel's Stranger Ideas', *Journal of the Royal Astronomical Society of Canada*, LXXV (1981), p. 46

Kent, Bill, 'Barnard's Wobble', *Swarthmore College Bulletin*, March 2001

Peperkamp, B., '"Mannekens in de maan" van Nicolaas Beets', *Nederlandse letterkunde*, IX/2 (July 2004), pp. 101–41

Phillips, Patricia, *The Scientific Lady: A Social History of Women's Scientific Interests, 1520–1918* (New York, 1990)

Racine, Jean, (1639–1699), 'Épigrammes, Sur l'Aspar de M. de Fontenelle: L'origine des Sifflets, St John Lucas, comp. (1879–1934)', *The Oxford Book of French Verse* (Oxford, 1920)

See, T. J., 'Recent Discoveries Respecting the Origin of the Universe', *Atlantic Monthly* (October 1897), pp. 484–92

——, 'Perturbations in the Motion of the Double Star 70 Ophiuchi = Sigma 2272', *Astronomical Journal*, xv (1895), p. 180

Sherrill, T. J., 'A Career of Controversy: The Anomaly of T. J. See', *Journal for the History of Astronomy*, xx (1999), p. 25

Smit, N. L., '"Een filosofisch geschriftje": Christiaan Huygens' gedachten over God in zijn Cosmotheoros en andere geschriften', Studium, *Tijdschrift voor wetenschapsgeschiedenis*, VII/1 (2014), pp. 1–18

Van Heertum, Cis, '"This is the Universe: Big Isn't It?" – Peter the Great and Christiaan Huygens', 2013, www.ritmanlibrary.com/2013/06/peter-the-great-and-christiaan-huygens

Wittenmyer, R. A., et al., 'Detection Limits from the McDonald Observatory Planet Search Program', *Astronomical Journal*, CXXXII (2006), p. 177

4 The Order of the Dolphin

Alexander, Amir, 'The Search for Extraterrestrial Intelligence: A Short History', *Cosmic Search Magazine*, www.planetary.org, accessed 2014

Baker, Andrew, 'The Search for Extra Terrestrial Intelligence', Lecture 14, Physics 343, Observational Radio Astronomy, 2014, Rutgers University

Basalla, George, *Civilized Life in the Universe: Scientists on Intelligent Extraterrestrials* (Oxford, 2005)

Burnell, Jocelyn Bell, 'Little Green Men, White Dwarfs or Pulsars?', *Cosmic Search Magazine*, I/1 (1979)

Drake, Frank, 'A Life with SETI', Stanford University Talks, 2010

—, 'Estimating the Chances of Life Out There', *Silicon Valley Astronomy Lectures*, 20 April 2005

—, and Dava Sobel, *Is Anyone Out There?* (New York, 1992)

Dyson, Freeman, *Disturbing the Universe* (New York, 1979)

Hewish, A., et al., 'Observation of a Rapidly Pulsating Radio Source', *Nature*, CCXVII (1968), pp. 709–13

Klaes, Larry, Paul Glister and Giuseppe Cocconi, 'SETI Pioneer', www.centauri-dreams.org, 16 December 2008

Marat, Balyshev, *Otto Ludwigovich Struve (1897–1963)* (Moscow, 2008)

Spangenburg, Ray, and Kit Moser, *Carl Sagan: A Biography* (Westport, CT, 2004)

Struve, Otto, 'Proposal for a Project of High-precision Stellar Radial Velocity Work', *The Observatory*, LXXII (1952), p. 199

Tarter, Jill, 'Beating the Odds', interview on www.seti.org, accessed March 2014

Weiss, Michael J., and Maria Wilhelm, 'Little Green Men to Earth: "Is Anybody Down There Listening?" Congress Debates the Answer', *People Magazine* (19 October 1981)

5 The Tenacious Inventor

Bathala, Nathalie, 'HAT-p-7b Confirmation and Many Great Things To Come', 5 August 2009, beyondthecradle.wordpress.com

Bhattacharjee, Yudhijit, 'Mr Borucki's Lonely Road to the Light', *Science*, CCCXL (2013), p. 542

Borucki, W. J., and A. L. Summers, 'The Photometric Method of Detecting Other Planetary Systems', *Icarus*, LVIII (1984), p. 121

——, R. L. McKenzie and C. P. McKay, 'Spectra of Simulated Lightning on Venus, Jupiter, and Titan', *Icarus*, LXIV (1985), p. 221

Burger, John Robert, *Human Memory Modeled with Standard Analog and Digital Circuits: Inspiration for Man-made Computers* (Hoboken, NJ, 2009)

Darwin, Charles, 'Letter to J. D. Hooker', 1 February 1871, Darwin manuscript collection, Cambridge University Library, 94:188–9

Emlen, Stephen T., et al., Cornell University Faculty Memorial Statement, 1971

Grant, Andrew, 'Planet Hunter', *Discover Magazine* (December 2012)

Rosenblatt, Frank, 'A Two-color Photometric Method for Detection of Extra-solar Planetary Systems', *Icarus*, XIV/1 (1971), pp. 71–93

Tappert, Chuck, *Dr Frank Rosenblatt, 1928–1971*, Pace University, DCS891B Research Seminar 2

Witze, Alexandra, 'Berkeley Releases Report on Astronomer Sexual-harassment Case', *Nature News* (19 December 2015), www.nature.com

Public Records Act documents, as posted on Geoffrey Marcy's public website, http://geoffreymarcy.com, accessed 26 September 2016

6 A Planet in Pegasus

Billings, Lee, *Five Billion Years of Solitude: The Search for Life Among the Stars* (New York, 2013)

Boss, A. P., 'Proximity of Jupiter-like Planets to Low Mass Stars', *Science*, CCLXVII (1995), pp. 360–62

Exalto, N., and H. Geurts, 'Het muzikale Doppler experiment van Buys Ballot', *Zenit* (1 October 1996)

Fellgett, P. B., *Optica Acta*, IX (1953)

Heidmann, J., and M. J. Klein, *Bioastronomy: The Search for Extraterrestrial Life* (New York, 1990)

Latham, David, 'The Unseen Companion of HD 114762', *New Astronomy Reviews*, LVI (2012), p. 16

Lemonick, Michael, *Mirror Earth* (London, 2012)
Mayor, M., and D. Queloz, 'A Jupiter-mass Companion to a Solar-type Star', *Nature*, CCCLXXVIII (1995), p. 355
—, 'From 51 Peg to Earth-type planets', *New Astronomy Reviews*, LVI (2012), p. 19
—, et al., '51 Pegasi', IAU *Circular*, VMCCLI (1995), p. 1, ed. B. G. Marsden
—, and P. Y. Frey, *New Worlds in the Cosmos: The Discovery of Exoplanets* (Cambridge, 2003)

7 The Hut in the Car Park

Boss, Alan, *The Crowded Universe: The Race to Find Life Beyond Earth* (New York, 2009)
Jaywardhana, Ray, *Strange New Worlds: The Search for Alien Planets and Life Beyond Our Solar System* (Princeton, NJ, 2011)
Sasselov, Dimitar D., '2008: What Have You Changed Your Mind About? Why?', www.edge.org/response-detail/10292

8 Goldilocks and the Red Dwarfs

Edge.org, 'Life – What a Concept!', interview with Dimitar Sasselov, www.edge.org, 27 August 2007
Exoplanet data explorer, http://exoplanets.org
Howell, Elizabeth, 'Are "Super-Earths" and "Habitable Zones" Misleading Terms?', www.astrobio.net, 26 May 2014
Locey, Bill, 'Sky's the Limit for Singing Astrophysicist', www.vcstar.com, 2012
'Meet Our Scientists – Elisa Quintana', on www.seti.org
National Science Foundation, press conference, 'Steven Vogt and Paul Butler Lead a Team that Discovered the First Potentially Habitable Exoplanet', http://nsf.gov/news, 2007
Overbye, Dennis, 'One Planet is Too Hot for Life, One May Be Just Right', *New York Times* (13 June 2007)
Quintana, E., et al., 'An Earth-sized Planet in the Habitable Zone of a Cool Star', *Science*, CCCXLIV (2014), p. 6181
Robertson, P., et al., 'Stellar Activity Masquerading as Planets in the Habitable Zone of the M Dwarf Gliese 581', *Science*, CCCXLV (2014), pp. 440–44

255

Swift, J., et al., 'Characterizing the Cool KOIS. IV. Kepler-32 as a
  Prototype for the Formation of Compact Planetary Systems
  throughout the Galaxy', *Astrophysical Journal*, DCCLXIV (2013),
  p. 105

Vogt, S. et al., 'The Lick-Carnegie Exoplanet Survey: A 3.1 M Planet
  in the Habitable Zone of the Nearby M3V Star Gliese 581',
  Astrophysical Journal, DCCXXIII/1 (2010), p. 954

Von Bloh, W., et al, 'The Habitability of Super-Earths in Gliese 581',
  Astronomy and Astrophysics, CDLXXVI/1 (2007)

Wall, Mike, 'Is Planet Gliese 581g Really the "First Potentially
  Habitable" Alien World?', www.space.com, 2012

9 Beer in Space

Deamer, David, 'Interstellar Cosmochemistry and Yellow Stuff
  from Outer Space', *Science 2.0* (www.science20.com), 23 April
  2009

Hagen, W., L. J. Allamandola and J. M. Greenberg, 'Interstellar
  Molecule Formation in Grain Mantles: The Laboratory Analog
  Experiments, Results and Implications', *Astrophysics and Space
  Science*, LXV (1979), p. 215

Tielens, A.G.G.M., and L. J. Allamandola, 'Absorption Features in
  the 5-8 Micron Spectra of Protostars', *Astrophysical Journal*,
  CCLXXXVII (1984), p. 697

10 The Space Rebels

Atreya, Sushil K., et al., 'Methane and Related Trace Species on Mars:
  Origin, Loss, Implications for Life, and Habitability', *Planetary
  and Space Science*, LV/3 (2007), p. 358

Bahcall, John N., 'Hearing before the Committee on Science, Space,
  and Technology', U.S. House of Representatives, in *Hubble Space
  Telescope Flaw*, One Hundred First Congress, second session,
  13 July 1990, p. 105

BBC Radio 4, 'Andre Geim Profile' by Helen Grady, first broadcast
  27 July 2013

Billings, Lee, "How NASA's Next Big Telescope Could Take Pictures
  of Another Earth", *Scientific American* (May 2016),
  www.scientificamerican.com

British Library, 'Alexander Fleming (1881–1955): A Noble Life in Science', www.bl.uk/onlinegallery

Carlson, K. B., 'Meet Astrophysicist Sara Seager, Canadian Genius', *The Globe and Mail* (25 September 2013)

Des Marais, D., et al., 'Remote Sensing of Planetary Properties and Biosignatures on Extrasolar Terrestrial Planets', *Astrobiology*, II (2002), p. 2

Ohlheiser, Abby, 'This Alien Hearing is the Best Thing Congress Has Done in Months', *Wired* (4 December 2013)

Seager, Sara, 'Written in the Stars', *Astrobiology*, XII (2012), p. 1

——, 'Sara's Story Collider Story', Facebook, 23 April 2014

——, 'Is There Life Out There? The Search for Habitable Exoplanets', www.saraseager.com, 2009

——, W. Bains and R. Hu, 'A Biomass-based Model to Estimate the Plausibility of Exoplanet Biosignature Gases', *Astrophysical Journal*, DCCLXXV (2013), p. 104

Werner, Debra, 'Mission To Search for Exoplanets One Star at a Time', spacenews.com, 18 April 2011

Yurchenko, S. N., et al., 'Spectrum of Hot Methane in Astronomical Objects Using a Comprehensive Computed Line List', *Proceedings of the National Academy of Sciences of the United States of America*, published online at www.pnas.org on 16 June 2014

11 The Planetary Circus

Güdel, M., et al., 'Astrophysical Conditions for Planetary Habitability', in *Protostars and Planets* VI (Tucson, AZ, 2014)

Maassen, Theo, *Functioneel Naakt* (2002)

Sagan, Carl, and Ann Druyan, *Pale Blue Dot: A Vision of the Human Future in Space* (New York, 1994)

The Wikipedia and SAO/NASA Astrophysics Data System websites were a useful starting point for much of the study of sources.

ACKNOWLEDGEMENTS

This book would not have been possible without the cooperation and support of many people. First of all, I would like to thank the planet hunters who made me so welcome at their places of work or spoke to me online. In no particular order, they are Tim Brown, Bill Borucki, Geoffrey Marcy, Seth Shostak, Jill Tarter, Michel Mayor, Didier Queloz, Lou Allamandola, Sara Seager and her research team, David Latham, John Johnson, Andrea Dupree, Andrew Walsh, Elisa Quintana, Jon Swift, Tim Lee, Francesco Palla, Robert Noyes, Stéphane Udry, Paul Kalas, Jean-Michel Desert and Mercedes Lopez Morales.

My thanks also to those who helped or advised me in other ways, including Ed van den Heuvel, Ineke Huysman of the Huygens ING, Henrik Beuther, Steve Croft, Teije de Jong, Martin Harwit, Marieke Baan, Govert Schilling, Jaap Vreeling, Carolyn Wever, Ton Raassen, Jan Krimp, David Cohen, Nuria Calvet and Mihkel Kama. I thank the Teylers Museum and the pupils who took part in the Planets teaching programme at the Stedelijk Gymnasium Haarlem for their help with such an inspiring project. My journey to meet the planet hunters was made possible by the Netherlands Research School for Astronomy.

Further thanks go to Koen Maaskant, for the clear illustrations of the planet measurement methods, to Michiel Haverlag, for his help in setting up the website www.planetenjagers.nl, and to Doris Zevenbergen, Ronit Palache and Mireille Berman, for their enthusiastic work in promoting the book.

Many thanks to Juliët Jonkers, who urged me to write, and to Katrijn van Hauwermeiren, my editor, for her dedication and expert advice. To Carsten Dominik, Bill Borucki, Sara Seager, Tim Brown, Seth Shostak, Ignas Snellen, David Garvelink, Job Breemer ter Stege, Juliët Jonkers, Koen Maaskant, Lex Kaper and Gerard 't Hooft: thanks for all your help in editing the text and sifting out factual errors. Gabrielle van Poll's quick and skilful work brought the planet hunters' words to life in a preliminary English translation. Dries Muus's painstaking editing made the book much more readable. My thanks also to Andy Brown, for his dedicated and enthusiastic work on the full English translation.

And, last but not least, my inexpressible thanks to Brechje, my first reader, for your ideas, your comments, your support and your love.

INDEX

51 Pegasi (star) 127, 139–40,
    143–5, 148, 150–51, 162–3, 168
51 Pegasi b (planet) 144, 148,
    150–52, 155, 163–4, 172, 180
61 Cygni (star) 72
67P/Churyumov-Gerasimenko
    (comet) 208
70 Ophiuchus (star) 73–4, 76

absorption line 135, 138, 141, 162,
    172, 215, 220, 224, 242
    see also spectral line
Adams, George 10
Allamandola, Lou 195, 198–9,
    200–201, 203–6, 208
Allen, Paul 99
Alpha Centauri (star) 57–9, 72,
    79, 221
Ames Research Center 195
amino acids 107, 203–5
Andriesse, C. D. 46–7
Apollo 71, 97, 102, 104, 175
    missions 99, 102, 195
Archimedes 43–4, 47, 132
Aristarchus 43
Aristotle 22–3
Armstrong, Neil 81, 102, 195
astrobiology 90, 206, 229–30, 242
    see also bioastronomy

astrochemistry 195, 201
astrometrists 71–2, 77, 80, 133
astrometry 71–2, 77–8, 242
Atacama Large Millimeter Array
    (ALMA) 208
Atchley, Dana 89

Bacon, Francis 35
Bahcall, John 217, 227
Barnard's Star 79–80, 94
Batalha, Natalie 124
Baur, Tom 164
Beets, Nicolaas 70
Bell, Jocelyn 92–3, 117
Bellarmine (cardinal) 37
Berkeley, University of California
    90, 96, 98–9, 128–30, 153, 169,
    187, 198, 239
Bessel, Friedrich Wilhelm 72
Billingham, John 94–6, 97–8, 142
binary star 72, 74, 76, 135, 141, 143,
    217, 233
bioastronomy 142, 242
biomarker 179, 206, 223–7,
    229–30, 234, 239, 242
Borucki, Bill 97, 102–10, 113–21,
    124–5, 130, 135, 156, 163–4,
    166, 171, 174, 176, 185, 189–90,
    205, 238

Boss, Alan 144, 146, 149, 219

Boyer, Stu 96

Boyle, Willard 247

Brahe, Tycho 28, 35, 45, 52, 54, 72, 239
see also Tycho

brown dwarf 78, 96, 137–8, 142, 145, 150, 158, 242

Brown, Tim 148, 150–51, 158–9, 165–70, 172, 174, 184, 215, 222

Bruno, Giordano 16–20, 26–8, 35–8, 40, 44–5, 63, 66–7, 71, 239

Bullock, Sandra 58, 223

Butler, Paul 139, 141, 151–3, 162, 168–9, 182–4, 219

Buys Ballot, Christophorus 134

California Institute of Technology (Caltech) 130, 187

Calvin, Melvin 90, 92

Campbell, Bruce 138, 143

carbon dioxide 105, 181, 203, 224

Carlin, George 12

Cash, Webster 222

Cassini, Giovanni Domenico (Jean-Dominique) 49–54, 56

Chandrasekhar, Subrahmanyan 88

Charbonneau, David 156–8, 165, 167–74, 180, 189, 212, 215, 238

Charge Coupled Device (CCD) 116–17, 120, 176

Clarke, Arthur C. 99, 101, 159

Clooney, George 58

CO₂ *see* carbon dioxide

Cocconi, Giuseppe 84–5, 101

Coleman, Sidney 247

Columbus, Christopher 152

comet 43, 59, 68, 72, 74, 193, 206–8, 224, 242

Conversations of the Plurality of Worlds (book) 64–5

Cook, James 52, 112

Cool Stars (conference) 145, 147

Copernicus, Nicolaus (Mikolaj Kopernik) 20, 66, 194

Cornell University 84, 95, 111

Cornell, Ezra 95

CoRoT (space telescope) 171, 180

COROT-7 b (planet) 180

cosmology 110, 130, 157

Cosmotheoros (book) 54, 60–61, 63–4, 100, 226

Curiosity Mars Rover 233

Darwin, Charles 17, 24, 106–7, 193, 205

De Witt, Johan and Cornelis 54

Deamer, David 205–6

Deming, Drake 173

Denver, John 165

Descartes, René 65

Dialogue Concerning the Two Chief World Systems (book) 36

DNA 107, 192, 204

Doppler, Christian 134

Doppler effect/method 89, 123, 133, 135, 138, 140, 148, 153, 158, 164, 166, 171, 185, 196, 220, 242

Drake Equation 86–7, 91, 95, 97, 109, 111, 125, 142, 225

Drake, Frank 83, 85–7, 90, 92, 94, 111, 125, 142, 184, 188, 197, 225, 234

Draper, Henry 246

Dupree, Andrea 147
Duquennois, Antoine 141
Dyson, Freeman 100

Earth-like planets 109, 113–15,
    164, 187, 191, 219–22, 232,
    243
Eddington, Arthur 160–61
Einstein, Albert 33–4, 75, 78, 98,
    160, 187, 215
El Niño 216
Enceladus (moon) 233
epicycles 22–4
ESA 233
Euclid 65
Europa (moon) 233
exomoon 232
ExoplanetSat 221, 223, 229
expansion of the universe 132

Fabrycky, Daniel 190
Fellget, Peter 141
Ferguson, James 67
Feynman, Richard P. 187
Fischer, Debra 139
Flash Gordon (TV series) 95
Fleming, Alexander 214
Fontenelle, Bernard le Bovier de
    52–3, 64–6, 80, 194
formaldehyde 197, 201
Formalhaut (star) 207–8
four elements 177–8, 180
Frankenstein (book) 106
Frederik Hendrik of Orange 48
Freedman, David 185
FRESIP 118, 163
    see also Kepler (space
    telescope)

Galilei, Galileo 24, 35–7, 47, 54,
    56, 61, 66, 146–7, 194, 233
Geim, André 215
geocentric world-view 36
George III, king of England 10–11
Gilliland, Ron 162–3, 166
GJ 1214 b (planet) 180
GJ 581 c (planet) 181–2
GJ 581 d (planet) 181–2, 184
GJ 581 g (planet) 182–4, 191
GJ 876 d (planet) 180
Goldilocks zone 179, 181–2,
    185–6, 190, 234–6
Grant, Andrew 68–9
Gravity (film) 58
Great Moon Hoax 70, 194
Great, Peter the 60
Green Bank (observatory) 85–7,
    89–91, 94, 97, 142
Greenberg, Mayo 199, 204
Güdel, Manuel 234–6

Halley, Edmund 245
Harvard University 76, 132,
    142–8, 156–7, 165, 167, 171,
    180, 210, 215, 229, 232, 238
Hawking, Stephen 157
HD 209458 (star) 166–72
Heinlein, Robert 159
Heintz, Wulff 79–80, 109
heliocentric world-view 23, 25–6,
    31, 36, 43, 45, 66, 147
Henry, Gregory 168–9
Herbig, George 131–2
Herschel (space telescope) 208
Herschel, Caroline 68
Herschel, John 68–70, 75–6, 80,
    194
Herschel, William 11, 67, 197

Herzberg, Gerhard 88
Herzing, Denise 83
Hewish, Antony 93
Hooft, Gerard 't 247
Hooker, Joseph 106
hot Jupiter 152, 155, 164, 172–3, 180, 243
Huang, Su-Shu 89
Hubble, Edwin 77, 132, 218
Hubble Space Telescope 58, 166–7, 172, 207, 217, 221, 233
Huygens, Christiaan 46–9, 50, 53–7, 59–65, 70, 75–6, 80, 100, 129, 226–7, 239
Huygens, Constantijn 46, 48, 54, 60
Huygens Sr, Constantijn 46–7
Huygens, Suzanna 46
hydrogen 108, 135, 138, 161, 196–7, 226–7

Independence Day (film) 81, 101
infrared 173, 197
    radiation 173, 197–8, 208
Institute for Advanced Study, Princeton *see* Princeton

Jacob, William 72–3, 78, 133
James Webb Space Telescope 221
Jansen, Sacharias 36
Johansson, Scarlett 99
Johnson, John 187, 189
Joy Division 93
Jupiter 11, 20–21, 28, 31, 34, 36–7, 41–2, 51, 59, 79, 89, 106, 108–9, 121, 126–7, 136–7, 144–6, 167–8, 178, 185, 187, 207, 224, 233, 235

Jupiter-like planets 112–13, 139, 145–6, 149, 164
Jupiter Icy Moon Explorer (JUICE) 233

Kaltenegger, Lisa 231
Kamp, Peter van de 77–80, 94, 96, 109–10, 133, 145, 183, 239
Kepler (space telescope) 118–21, 124–5, 136, 163–4, 166, 171, 176, 180, 185–91, 220–21, 223, 234, 238–9
Kepler Orrery 190
Kepler-10 b (planet) 186
Kepler-186 f (planet) 190
Kepler-22 b (planet) 186
Kepler-32 (star, planetary system) 189
Kepler-62 e (planet) 186
Kepler-62 f (planet) 186
Kepler-78 b (planet) 186
Kepler, Johannes 24, 28, 30–35, 64, 112, 133, 135, 144, 194, 239
King, Larry 83
Koch, David 117–18
Koestler, Arthur 28, 32
Kuiper Airborne Observatory 202

Late Heavy Bombardment 207–8
Latham, David 142, 166–7, 169
Leclerc, Sébastien 51
Lederberg, Joshua 90
Lederman, Leon 34
Lee, Tim 230
Lemonick, Michael 152
Letterman, David 153
Lévi-Strauss, Claude 14
LGM-1 93

Lick Observatory 119, 130, 139
Lilly, John 90, 92
Lincoln, Abraham 223
Lipperhey, Johannes 36
Locke, Richard 69, 70
Lockwood, Robert P. 17, 18
Lord of the Rings, The (films) 207
Louis xiv, king of France 49
Lowell, Percival 76, 80
Luther, Martin 25

Maassen, Theo 237
Madhusudhan, Nikku 228
Mallén-Ornelas, Gabriela 216–17
Manhattan Project 84
Marcy, Geoffrey 128–33, 135–9,
      141, 143, 148, 150–54, 156, 162,
      164, 168–70, 174, 177, 181–2,
      196, 223, 239–40
Mars 11, 21–3, 31, 33, 41–2, 45,
      50–54, 59, 76–7, 80, 142, 170,
      178–9, 198, 224–5, 236, 239
Massachusetts Institute of
      Technology (MIT) 210, 214,
      219–21, 227, 229–30, 233
Matrix, The (film) 211
Mayor, Michel 140, 143, 147–53,
      162–4, 166, 171, 180, 196
Mazeh, Tsevi 166
McClintock, Barbara 218
Men in Black (films) 12
Mercury 11, 21, 41–2, 112, 144,
      179
Mersenne, Marin 47
meteorite 92, 103, 205–8, 243
methane 107, 224–6, 230
Mihalas, Dimitri 160
Miller, Stanley 107–8, 205
Mocenigo, Giovanni 19

moon 10, 36–7, 46, 48, 51, 67–8,
      72, 129, 146, 190, 232–3
Moon, the 21, 25, 29, 36, 56, 61,
      64, 68–71, 81, 84, 94, 98, 102,
      104–5, 159, 161, 195, 198,
      207, 224, 236
Moore, Gordon 99
Morrison, Dave 116
Morrison, Philip 84–5, 90
Moulton, Forest Ray 74
Mysterium cosmographicum
      (book) 31

NASA 94–5, 98, 104–5, 109, 117–18,
      120–21, 163, 171, 175, 183,
      185–6, 198, 206, 221–3
Nature (journal) 85, 93, 142,
      145–6, 151
Neptune 11, 21, 41–2, 57, 59,
      180–81, 224
Neutron star 93, 117
Newton, Isaac 34–6, 46, 165
Nobel prize 88, 90, 93, 128, 131,
      153, 215
Noyes, Bob 148, 151, 157–8, 163

Observatoire de Haute-Provençe
      140
Oliver, Bernard 89, 94, 98
Order of the Dolphin 81–101, 110,
      136, 142
Origin of Species, The (book) 24
orrery 10–11, 190
oxygen 135, 178–9, 199, 208,
      224–5, 227
ozone 106, 179, 224–5, 235

Palla, Francesco 147–9
parallax 43–5, 50, 52, 55, 57, 72, 88

Pauling, Linus 187
PDP-8/s (computer) 96, 99
Peter the Great 60
photometry 166
Pimentel, George 198–9
planetary atmosphere 106, 109,
    172–3, 179, 182, 222, 226, 230
planetary transit 112–13, 120–21,
    168, 171, 216–17, 220
planetoid 67
Pluto 11, 41, 59, 151, 240
Poe, Edgar Allan 70
Princeton 215–18, 232
Project Cyclops 94
Project Ozma 85, 92–3
Protostars and Planets
    (conference) 231
Proxmire, William 97
Ptolemy 23–4, 31, 45
pulsar 93, 117–18

Queloz, Didier 140, 142–7, 149–53,
    162, 171, 180, 238
Quintana, Elisa 175–7, 186,
    190–92

Racine, Jean 65
radio 12, 77, 83, 85–7, 91, 93–4,
    100–101, 104, 106, 197, 208,
    225
    astronomy 197
    signals 81–2, 85–6, 90, 92–3,
        96, 98, 100
    telescope 83–6, 92, 100, 192,
        239
    waves 85, 91, 94, 197
red dwarf 188–9, 191, 220, 235,
    243
retrograde motion 22

Revolutionibus (book) 24–6, 66
Rheticus 24
Richer, Jean 49–54, 56
Richter, Charles 218
Ride, Sally 176
Romein, Annie 63
Romein, Jan 63
Rosenblatt, Frank 110–14, 156
Rosetta (spacecraft) 208
Rowling, J. K. 178
Rudolf II, Holy Roman Emperor
    30

Sagan, Carl 90, 97, 111, 116, 118,
    130, 240
Sasselov, Dimitar 171, 180–81,
    215, 230
Saturn 11, 20–21, 28, 31, 41, 42, 46,
    48, 59, 68, 106, 129, 168, 178,
    224, 232–3
Savio, Mario 139
Science (journal) 144, 184, 191–2
science fiction 60, 64, 66, 70, 77,
    84, 90, 95, 98–9, 110, 194
Seager's Equation 234
Seager, Sara 157, 171–4, 210–23,
    225–6, 228–30, 238
Search for Extraterrestrial
    Intelligence (SETI) 81–5, 87,
    90, 92–101, 104, 109–11, 142,
    175, 186, 192, 197, 206, 225,
    243
See, Thomas Jefferson 73–6, 78
Shelley, Mary 106
Shostak, Seth 82–3, 94, 99, 101,
    192, 239
Sirius (star) 20, 55–7, 59, 60,
    62–3
Snellen, Ignas 230

Snyder, Lewis 197–8
Somnium (book) 64
Song of Ice and Fire, A (book) 30, 178
Space Shuttle 129
spectral line 196–7, 201, 209, 243
spectrograph 135, 137–9, 141, 143, 153, 155, 162, 163–4, 192, 196, 198, 201–2, 221, 223, 224, 243
spectroscopy 131–3, 135, 158, 164, 196, 198
spectrum 131, 135, 138, 141, 162, 172–3, 184, 194, 196, 201, 208, 220, 223–4, 229–30, 243
Sputnik 159
Stampioen, Jan 46–7
star formation 131, 197, 209
Star Trek (TV series and films) 77, 80, 113, 120
Star Wars (films) 77, 223
STARE (Stellar Astrophysics and Research on Exoplanets) 163
Starshade 222–4
stellar pulsations 158, 163, 167
Struve, Otto 87–9, 97, 103, 109, 111–12, 135, 152, 156
Summers, Audrey 114–15
super-Earth 180–81, 187, 219, 244
Swift, Jon 187–9, 191

Tarter, Jill 95–101, 110, 116, 118, 130, 137, 142, 191, 239
Terrestrial Planet Finder 222–3
Teylers Museum 9, 11–13, 190, 241
Tielens, Xander 201, 208

Titan 46, 48, 109, 129, 232
transit 52, 111–16, 120–22, 155, 158, 164–73, 187–9, 215–16, 220, 238
transit method 89, 111, 135, 158, 180, 215, 244
see also planetary transit
Transiting Exoplanet Survey Satellite (TESS) 220–21, 223
Tycho 28–33, 35, 45, 52, 54, 72, 239

ultraviolet 199, 201
radiation 235
University of Amsterdam 17
University of Cambridge 92, 143, 228
University of Geneva 140
University of Nijmegen 215
University of Oxford 19, 26
Uraniborg 29–30
Uranus 11, 21, 41–2, 59, 67–8, 181, 224
Urey, Harold 107–8, 205
Ursus 31–2

Venus 11, 20–21, 36, 41–2, 52, 106, 109, 112, 179, 181, 235–6
Very Large Telescope (VLT) 13, 153
Vogt, Steven 139, 182–4, 209
Vogt, Zarmina 182, 184
Voltaire 53
Voyager 1 106, 240
Vulcan Telescope 120, 130, 171, 176

Walker, Gordon 138, 146

Walsh, Andrew 202–3
War of the Worlds (book,
    radio play, film) 12, 77
Wells, H. G. 77
William iii 60
Witt, Cornelis de 54
Witt, Johan de 54

X-Files, The (tv series) 92
X-ray radiation 235

Yerkes (observatory) 88, 103

Zarmina's World 182, 184,
    209